Introduction to Applied Optics for Engineers

Introduction to Applied Optics for Engineers

F. Paul Carlson

Department of Electrical Engineering
University of Washington
Seattle, Washington

ACADEMIC PRESS New York San Francisco London 1977

A Subsidiary of Harcourt Brace Jovanovich, Publishers

ACADEMIC PRESS, INC.
111 Fifth Avenue, New York, New York 10003

United Kingdom Edition published by
ACADEMIC PRESS, INC. (LONDON) LTD.
24/28 Oval Road, London NW1

Library of Congress Cataloging in Publication Data

Carlson, F. Paul, Date
Introduction to applied optics for engineers.

Includes bibliographical references and index.
1. Optics. I. Title.
TA1520.C37 621.36 75-3575
ISBN 0-12-160050-5

PRINTED IN THE UNITED STATES OF AMERICA

To Judy
and the children

CONTENTS

PREFACE

This book is designed to provide a broad base of material in applied optics for students and engineers who have a traditional background in electromagnetic wave theory. Often students are not exposed to the extension of electromagnetic wave concepts, to Kirchhoff-type integral expressions, or to the relation between wave concepts and optics. Traditionally, optics books assume an adequate background on the part of the reader in wave propagation, and minimize the presentation of the transitional material from simple wave concepts to integral equations. Thus a good understanding of the underlying wave theory of applied optics is not provided.

Once the basic integral expression describing the radiated fields is derived from Maxwell's equations, it is used to develop the laws and principles of optics and optical devices from the wave-picture point of view. This development enables the reader to understand the origin of the basic laws of optics, and enables the development of solutions for those cases where first-order approximations fail. The full-wave approach makes the presentation of diffraction theory much easier.

The book then goes on to describe the ideas of modern coherent optical data processing with examples taken from current research work. Several examples from bioengineering-related research are presented with suggestions for further work by the interested researcher.

Through this kind of development advanced seniors and first-year graduate students can obtain a grasp of the evolution and usefulness of optical devices such as lenses, wedges, arrays, and other processor elements. Design constraints and questions of physical realizability follow quite naturally from the limits imposed on specific mathematical approximations. This method of developing design limits from the approximation development limits the ambiguity associated with ad hoc developments.

With this early foundation centered on the Rayleigh–Sommerfeld integral equation, the descriptions of optical processor systems, interferometers, lasers, holography, and other modern devices follow an evolutionary pattern of a mathematical description, and then go on to physical realization. In addition, the integral equation method is used as a basis for introducing and examining the ideas of partial coherence theory which are of current importance in the design of new detection devices and systems. Lastly, an overview of scattering theories is presented. The section on scattering theory begins with some simple observations of everyday phenomena and then proceeds to compare various current theories and research. The final section on multiple scattering provides a connection between the phenomenological and analytical approaches. Where possible, current and new results are presented to show how much optics, and particularly coherent optics, have developed in recent years. It seems that this is just a beginning in the quest to apply this technology.

This book is based on a set of course notes used in a one-quarter course in applied optics in the Department of Electrical Engineering at the University of Washington. In order to present this material in forty lectures it is necessary to keep a vigorous pace, particularly in the first four chapters. The problems are primarily designed to involve the student in understanding the text material, hence there are many derivations which are meant to be pedagogical. The style by which ideas are developed assumes that all the problems are worked. For some topics, the related ideas and developments occur in later chapters. This technique allows a smoother development of the basic ideas, and then provides a broader base for understanding actual systems and examples.

ACKNOWLEDGMENTS

I appreciate and acknowledge the comments and encouragement of Drs. Akira Ishimaru, David C. Auth, J. David Heywood, Sinclair Yee, Laurel J. Lewis, John L. Bjorkstam, James O. Raney and many others too numerous to list here. A special thank you is due to Mrs. E. Flewelling for her willingness and care in typing most of the manuscript. Finally, I am indebted most to my wife Judy, whose patience and encouragement was so helpful. The financial support of both the Office of Naval Research (Code 430C) and the National Institute of General Medical Sciences is acknowledged, with much appreciation, for much of the work reported on in the last six chapters.

Chapter I

AN INTRODUCTION TO PHYSICAL OPTICS

Introduction

In this chapter the basic equations of physical optics are derived, starting with the differential form of Maxwell's equations. The wave treatment approach to optics leads quite nicely to the simpler geometrical (ray) optics approximation, and also shows the correspondence with the traditional plane-wave approach, which retains the important diffraction effects. This derivation enables the reader to understand the limits of the two approximations and use them appropriately.

Propagation in Free Space

To understand the physical observations encountered in radiating electromagnetic systems, such as optical systems, a convenient starting point is afforded through Maxwell's equations [1, 2]. These equations consist of four first-order differential equations. Together with the constitutive relations, a complete set is obtained, having a simple and tractable form. Solutions of these equations can be obtained when the boundary conditions are specified.

The method of solution using these four first-order equations is difficult because the variables are mixed. One approach to simplifying these equations is to eliminate the mixed-variable character by increasing the order. This procedure produces the wave equation which can be solved by simpler techniques.

The procedure starts with the time-dependent form of Maxwell's equations

$$\nabla \times \mathbf{E} = -\partial \mathbf{B}/\partial t \tag{1-1a}$$

$$\nabla \times \mathbf{H} = \mathbf{J} + \partial D/\partial t \tag{1-1b}$$

$$\nabla \cdot \mathbf{D} = \varrho \tag{1-1c}$$

$$\nabla \cdot \mathbf{B} = 0 \tag{1-1d}$$

and the constitutive relations [1, 2]

$$\mathbf{D} = \varepsilon \mathbf{E} \tag{1-2a}$$

$$\mathbf{B} = \mu \mathbf{H} \tag{1-2b}$$

where ε and μ are the permittivity and permeability of the media. If we consider that the media in which the waves are propagating are isotropic and homogeneous, then the permittivity and permeability become simple scalar functions, and Eqs. (1-1) can be simplified. Furthermore, if the region is source free, then ϱ and $\mathbf{J}$ are zero, and Eqs. (1-1) become

$$\nabla \times \mathbf{E} = -\mu(\partial \mathbf{H}/\partial t) \tag{1-3a}$$

$$\nabla \times \mathbf{H} = \varepsilon(\partial \mathbf{E}/\partial t) \tag{1-3b}$$

$$\nabla \cdot \mathbf{E} = 0 \tag{1-3c}$$

$$\nabla \cdot \mathbf{B} = 0 \tag{1-3d}$$

Restricting consideration to only time-harmonic fields of the form $e^{-j\omega t}$ in the steady-state region, the time-differential terms further reduce to algebraic form

$$\nabla \times \mathbf{E} = j\omega\mu \mathbf{H} \tag{1-4a}$$

$$\nabla \times \mathbf{H} = -j\omega\varepsilon \mathbf{E} \tag{1-4b}$$

$$\nabla \cdot \mathbf{E} = 0 \tag{1-4c}$$

$$\nabla \cdot \mathbf{H} = 0 \tag{1-4d}$$

where ω is the radian frequency of the waves or 2π times the frequency f in hertz.

Equations (1-4) still involve mixed variables and can be further reduced to a more tractable form by taking the curl either of Eq. (1-4a) or of (1-4b) and using the vector identity

$$\nabla \times \nabla \times \mathbf{F} = \nabla(\nabla \cdot \mathbf{F}) - \nabla^2 \mathbf{F} \tag{1-5}$$

To illustrate, the curl of Eq. (1-4a) is

$$\nabla \times \nabla \times \mathbf{E} = j\omega\mu(\nabla \times \mathbf{H}) \tag{1-6}$$

Substituting Eq. (1-4b) in the right-hand side leaves

$$\nabla \times \nabla \times \mathbf{E} = j\omega\mu(-j\omega\varepsilon\mathbf{E}) = \omega^2\mu\varepsilon\mathbf{E} \tag{1-7}$$

Since the divergence terms in Eqs. (1-4c) and (1-4d) are zero, Eq. (1-7) is simplified by the substitution of Eq. (1-5) to obtain

$$\nabla^2\mathbf{E} + \omega^2\mu\varepsilon\mathbf{E} = 0 \tag{1-8}$$

Equation (1-8) is known as the wave equation for the electric field. The derivation of a similar equation for the magnetic field **H** is included in the problem set at the end of this chapter. The vector form of Eq. (1-8) is more complicated than need be. In many cases a relation involving only a single component will suffice.

To obtain a complete solution to wave equation, appropriate boundary conditions are required corresponding to the physical structure. A common case is represented by zero tangential fields at perfect conductors and a radiation condition [2, 3] which requires the field to be zero at infinity. Basically, the wave equation describes how fields exist in time and space. The coupling through the two terms in the equation indicates that traveling or propagating waves will exist in the region. To illustrate, a plane-wave case will be considered.

Plane-Wave Propagation

Consider an infinite half space that has only the boundary condition defined by the radiation condition [2, 3], which arises from a constraint that finite energy exists at infinity. Using this condition, the solutions are simplified. In addition, consider that only a forward traveling wave exists. The explicit form of this assumption will follow in subsequent equations.

Throughout this treatment we have been considering macroscopic descriptions of the media by using ε and μ. Conventionally, however, the permittivity ε is replaced by the index of refraction η, which is

$$\eta = \sqrt{\varepsilon/\varepsilon_0} \tag{1-9}$$

where ε_0 is the permittivity of free space. The index of refraction basically enables a description of how the phase of a wave changes as a function of a relative or normalized permittivity ε_r, where

$$\varepsilon = \varepsilon_r \varepsilon_0 \tag{1-10}$$

One additional parameter is the notion of propagation vector **k**, where k^2 is defined as [1, 2]

$$k^2 = \omega^2 \mu \varepsilon \tag{1-11}$$

The vector direction is associated with the direction that the wave front propagates. In terms of free-space conditions, this can be written

$$k^2 = k_0^2 \eta^2 \tag{1-12}$$

Thus the wave equation of Eq. (1-8) reduces for one component to

$$(\nabla^2 + k_0^2 \eta^2) u = 0 \tag{1-13}$$

where u represents any of the three possible components E_x, E_y, or E_z. It should be noted that the Laplacian ∇^2 in the rectangular coordinates becomes

$$\nabla^2 = \partial^2/\partial x^2 + \partial^2/\partial y^2 + \partial^2/\partial z^2 \tag{1-14}$$

If a simple one-dimensional problem is considered, Eq. (1-13) reduces to a simple form. The solution can be provided by an educated guess, or by noting that the form of the wave equation is like the classic harmonic function, leaving

$$u(x) = A e^{\pm jkx} \tag{1-15}$$

The amplitude of the wave is A, and $\pm jkx$ is the phase of the wave. Substitution of this result shows that it is indeed a solution. The plus/minus sign is used to indicate either forward or reverse waves. Remembering that an $e^{-j\omega t}$ time dependence was assumed, e^{+jkx} corresponds to a forward traveling wave.

The solution represented by Eq. (1-15) is known as the plane-wave solution. It is used extensively in the literature, primarily because most other complicated wave shapes can be decomposed into a sum of plane waves. It is called a plane wave because the phase fronts have no curvature or are said to be flat and constant in directions transverse to the direction of propagation. The wave vector **k** is also obtained quite simply from the gradient of kx, leaving $\hat{\imath}k$. Further, plane waves arise from sources that are very far away, thus satisfying the constraint of a source-free region. Because of the simple form associated with a plane-wave result, it will be used extensively in what follows.

Geometrical Optics

The plane-wave result, Eq. (1-15), suggests that a wave, satisfying the reduced scalar wave equation, can be represented by a simplified form having separable amplitude and phase terms of the form [3, 4]

$$u = Ae^{iS} \tag{1-16}$$

where A is the amplitude of the wave and S is the position-dependent phase term. Note that in the special case of Eq. (1-15), S is the scalar product of the propagation vector $\mathbf{k}$ and the position vector $\mathbf{x}$. This representation suggests that the notion of phase fronts moving through space may adequately describe the field in a particular region. Noting that surfaces can generally be described by their normals, then surfaces of constant phase, with corresponding normals, may have significant meaning. Specifically, it is found that the normal is aligned with the general propagation vector $\mathbf{k}$, as shown in Fig. 1-1.

In optics, particularly, these normals are referred to as rays. If the amplitude of the field A associated with a particular phase surface does not change appreciably over some appropriate distance of propagation, then the ray representation may embody most of the propagation phenomena, particularly in the short wavelength cases. That is, the wavelength is very

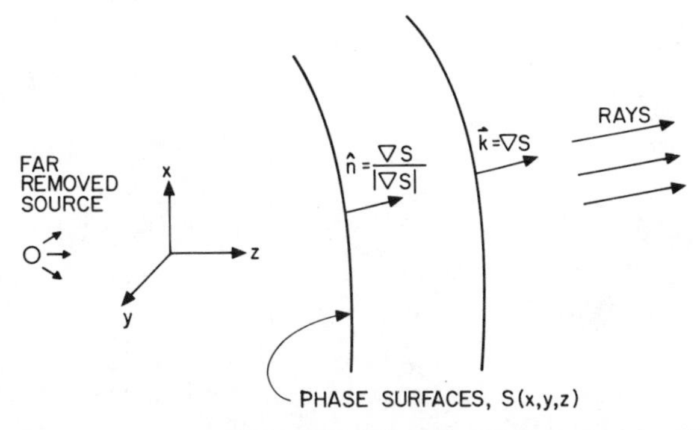

Fig. 1-1. Representation depicting equivalence of plane-wave models with large phase-front curvature and the ray-optic picture.

much smaller than the geometric constructs (e.g., lens sizes, focal lengths, turbulent eddy sizes) in the region. From this notion geometric optics can be defined as the limiting theory that describes the transport of energy in ray bundles. The easiest association with ray bundles is in the simple pictures describing lenses, wedges, and other ideas like Snell's law, which are first encountered in sophomore physics books. The notion of equivalence in these ideas is shown in Fig. 1-1.

In homogeneous media, these rays travel in straight lines, independent of each other. In inhomogeneous media, where the index of refraction is position dependent, the ray paths will be shown to be governed by a simple integral of the index of refraction over the traversed path. The basic law associated with this integral is Fermat's principle. One important result of this principle is that the paths may not necessarily be straight lines.

It can be shown how geometric optics follows from electromagnetic theory. To do so, use the plane-wave case governed by the homogeneous scalar wave equation, Eq. (1-13).

First, map $u(x, y, z)$ into the amplitude and phase space described by Eq. (1-16) [5]. Explicitly, divide Eq. (1-13) through by u and recognize that the first term can be related to the differential of $\log u$ as follows:

$$\nabla^2 u/u + k_0{}^2\eta^2 = \nabla^2(\log u) + (\nabla \log u)^2 + k_0{}^2\eta^2 = 0 \tag{1-17}$$

If Eq. (1-16) is substituted, Eq. (1-17) becomes

$$\nabla^2(\log A) + (\nabla \log A)^2 - (\nabla S)^2 + k_0{}^2\eta^2 + i(\nabla^2 S + 2\nabla \log A \cdot \nabla S) = 0 \tag{1-18}$$

This complex equation can be separated into two independent equations by noting that the real and imaginary parts must each be zero. Thus the separated equations become

$$\nabla^2(\log A) + (\nabla \log A)^2 - (\nabla S)^2 + k_0{}^2\eta^2 = 0 \tag{1-19a}$$

and

$$\nabla^2 S + 2\nabla \log A \cdot \nabla S = 0 \tag{1-19b}$$

The mathematical implementation of the short-wavelength approximation follows by dropping certain terms using order of magnitude arguments. This will produce a shorthand equation that describes the case of geometric optics.

Without much loss of generality, the analysis can be restricted to the case where amplitude changes in the medium can occur only over distances on the order of scale size changes, such as a lens size, an inhomogeneity length, boundary size, etc. Denoting this scale by l, the short-wavelength approximation is represented by writing $l \gg \lambda$.

For the plane-wave representation, the gradient of S, which corresponds to the phase-surface normal, varies as the propagation vector **k**. Further, Eq. (1-11), which defines the magnitude of **k**, can be rewritten in terms of wavelength as

$$k = 2\pi/\lambda \tag{1-20}$$

Using this expression for k, the third term in Eq. (1-19a) can be written as

$$(\nabla S)^2 \sim (1/\lambda)^2 \tag{1-21}$$

The first two terms of Eq. (1-19a) can be reduced to

$$\nabla^2 A/A = \nabla^2(\log A) + (\nabla \log A)^2 \tag{1-22}$$

using the arguments leading to Eq. (1-17). However, since amplitude changes only occur over distances of the medium scale size, Eq. (1-22) varies like inverse length squared,

$$\nabla^2 A/A \sim (1/l)^2 \tag{1-23}$$

Since the last term in Eq. (1-19a) also varies like the inverse wavelength and since

$$(1/l)^2 \ll (1/\lambda)^2 \tag{1-24}$$

Eq. (1-19a) can be reduced to

$$(\nabla S)^2 = (k_0\eta)^2 \tag{1-25}$$

which is classically called the eiconal equation. Thus, the eiconal equation describes the phase-surface gradients or unnormalized surface normals in terms of the medium propagation vector. In this case the eiconal function S is the function describing the phase surface. The gradient of S leads to the notion of a ray traveling perpendicular to the surface [3].

Thus, solutions of the differential equation, Eq. (1-25), gave the wave fronts associated with a geometrical-optics representation of propagation. This is a useful concept for many cases. It has limitations, however, arising from the contributions of the dropped term $\nabla^2 A/A$. Physically, this term corresponds to the bending or curvature of the waves by medium objects. The description of the phenomenon of bending of waves around obstacles is given by diffraction theory.

Hence, geometrical-optics solutions do not account for diffraction effects. These diffraction effects are usually restricted to a small region close to the forward direction of propagation, but they can be the dominant physical effect in many optical systems. This is referred to as a diffraction-limited system. Thus, as long as variations in amplitude are small, the so-called ray-optic method can be used. This is particularly true in first-order effects, such as placing lenses or calculating first-order reflection angles, etc.

It will, in fact, be shown in Chapter III that the simple lens law of thin lens optics

$$1/o + 1/i = 1/f \tag{1-26}$$

is still meaningful in diffraction-limited optical systems. There, however, this imaging condition will be derived from minimum phase-error considerations.

Fermat's Principle

Solutions of the eiconal equation are quite simple in homogeneous media. Basically, S becomes the scalar product of the propagation vector and the propagation distance, yielding the electrical phase length of the medium. Thus, interference and focusing have to do with evaluation of the total phase length of rays and noting differences that are multiples of π, etc. For example, interference is defined by the condition that two rays will destructively interfere if the phase difference is an odd multiple of π. Constructive interference corresponds to phase differences represented by multiples of 2π.

When the medium is inhomogeneous, the problem is more complex, and it is necessary to examine the changes in ∇S along a path. To see this, consider the projection of ∇S in the direction of the path increment $d\mathbf{l}$, i.e.,

$$\nabla S \cdot d\mathbf{l} = (\mathbf{k}_0 \eta) \cdot d\mathbf{l} \tag{1-27}$$

This is simply the change in phase in the direction of the path. From vector calculus, this is known as the directional derivative of S, dS, or

$$\nabla S \cdot d\mathbf{l} = dS \tag{1-28}$$

Therefore, if the total phase change along the path is desired, the changes represented by Eq. (1-28) are simply summed, giving

$$S = \int_0^L \mathbf{k}_0 \eta \cdot d\mathbf{l} \tag{1-29}$$

which is often written simply as

$$S = k_0 \int_0^L \eta \, dl \tag{1-30}$$

This is the mathematical statement of Fermat's principle. The question of what path the ray follows in propagating from P_1 to P_2 is determined by finding which path renders the time of travel from P_1 to P_2 stationary. In some cases this has inappropriately been extended to a least-time statement. It has been shown [3] that the stationary constraint may be neither a maximum or minimum.

Physically, the result of Fermat's principle is that paths are no longer necessarily straight, but rather are determined by the functional form of the index of refraction $\eta(x, y, z)$. Examples of where variable indices are important occur in gas lenses [6, 7], some fibers [8, 9], and in some plasma systems [10] such as rocket plumes.

Integral Relation for the Field

a. *Heuristic Approach*

At this point it is desirable to develop a representation for the fields at points in a remote aperture that are excited by some exciting aperture. Some of the facts known about the system, shown in Fig. 1-2, are that fields at S_1 will propagate to S_2. These fields satisfy the wave equation, Eq. (1-8), and each point in S_1 can be considered as a separate secondary point-source-like radiator. This last statement follows from Huygens' principle [11]. It also seems quite obvious that at every point (x', y', z') some contribution from each point (x, y, z) over S_1 must be summed to find the field at (x', y', z'). Lastly, each point in S_1 will radiate with some charac-

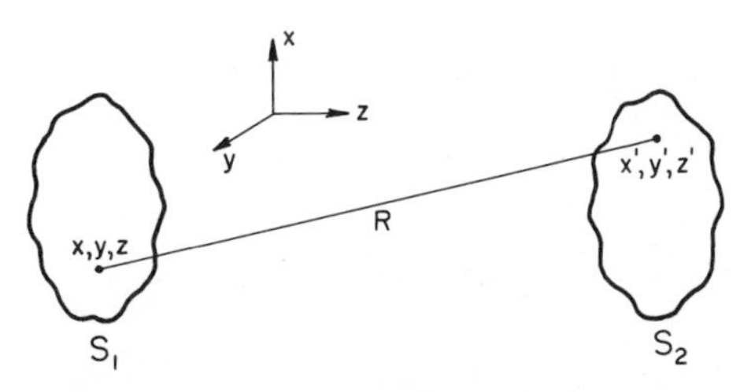

Fig. 1-2. General coordinate systems for radiating S_1 and receiving S_2 apertures.

teristic form to the points in S_2. This form is represented by the function† $K(x, y, z; x', y', z')$. Thus, if the field $u(x, y, z)$ is multiplied by $K(\ )$ and then summed over the whole aperture, the field $u_2(x', y', z')$ over S_2 is obtained, except for some unspecified constants. Mathematically, this can be expressed as

$$u_2(x', y', z') = (\ ) \iint_{S_1} u(x, y, z) K(x, y, z, x', y', z')\, dS_1 \qquad (1\text{-}31)$$

which is an integral equation relating the fields of S_1 and S_2. It remains to find the form of $K(\ )$ and the appropriate constants in front of the integral. For now, however, the form of our desired answer is given by Eq. (1-31). Additionally, one should note that Eq. (1-31) is consistent with the superposition principle and explicitly represents the field by a linear relationship.

b. *Analytic Approach*

The wave equations, Eq. (1-8) and Eq. (1-13), are certainly general enough that their form can be examined in some detail. These equations are reduced wave equations. A reduced wave equation, however, is a special form of a more general inhomogeneous wave equation, the solution of which is called a Green's function. The more general wave equation is

$$(\nabla^2 + k^2) G(x, y, z) = -\delta(x - x')\, \delta(y - y')\, \delta(z - z') \qquad (1\text{-}32)$$

where $\delta(\ )$ is the Dirac delta function‡ [12, 13]. When this equation corresponds to a physical radiating system, the right-hand side corresponds to a source located at a point (x', y', z'). Further, the constants have all been incorporated into $G(x, y, z)$ to match the inverse length dimensions of $\delta(\ )$.

In this section a very standard method will be employed to solve Eq. (1-32).§ This method is often used in electromagnetics, quantum me-

† In free space the function $K(\ )$ turns out to be e^{ikR}/R, where R is defined as the distance between (x, y, z) and (x', y', z'). $K(\ )$ is also called a propagator function.

‡ The Dirac delta function is defined by what it does to other mathematical functions, i.e.,

$$\delta(x - x') = 0 \quad \text{if } x \neq x', \quad \int_{-\infty}^{\infty} \delta(x - x')\, dx = 1, \quad \text{and} \quad \int_{-\infty}^{\infty} f(x)\, \delta(x - x')\, dx = f(x')$$

§ This ploy is the central manipulation of the Sturm–Liouville proof of orthogonality [14].

chanics, and other studies involving second-order differential equations. First, Eq. (1-13) is multiplied by G and Eq. (1-32) by u. Then the two equations are subtracted, yielding

$$G(\nabla^2 u + k^2 u) - u(\nabla^2 G + k^2 G) = +u\,\delta(x - x')\,\delta(y - y')\,\delta(z - z') \tag{1-33}$$

The left-hand side becomes[†]

$$G\nabla^2 u - u\nabla^2 G = +u\,\delta(r - r') \tag{1-34}$$

which can be reduced to

$$\nabla \cdot (G\nabla u - u\nabla G) = +u\,\delta(r - r') \tag{1-35}$$

using the identity [14]

$$\nabla \cdot (\phi\nabla\psi - \psi\nabla\phi) = \phi\nabla^2\psi - \psi\nabla^2\phi \tag{1-36}$$

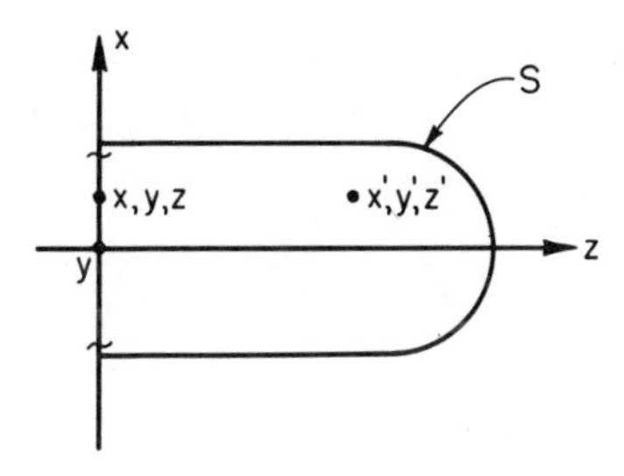

Fig. 1-3. Right semi-infinite half space for the surface integration.

To find the field $u(x', y', z')$ in the region of concern, namely, the right semi-infinite half space shown in Fig. 1-3, Eq. (1-35) is integrated over the volume bounded by S. The divergence theorem [14] is used to obtain[‡]

$$\oint\!\!\oint (G\,\nabla u - u\,\nabla G) \cdot \hat{n}\,dS = u(r') \tag{1-37}$$

The $\oint\!\!\oint$ indicates a closed surface bounding the volume in question.

[†] The use of r here is a shorthand vector notation meaning the ordered triplet.

[‡] Remember that the integral of the right-hand side is performed using the definition of a delta function, i.e.,

$$\int_{-\infty}^{\infty} f(x)\,\delta(x - x')\,dx = f(x')$$

This result indicates that the field at any point inside the region is determined by the field on the surface of the bounding region. Thus, all that is needed is the field description on the surface. The description of the surface fields will be broken into two parts—that which is on the xy plane and that which is on the hemispherical surface. Since the radius of the hemisphere can be very large, the fields on the surface must tend to zero as the radius increases in order that the radiation condition is satisfied. Similarly, since solutions for G obey a similar auxiliary or boundary condition, all the surface integral terms on the hemisphere must vanish.

To complete the specification of the boundary conditions on G, three types of conditions must be considered in general—the Dirichlet, Neumann, and mixed; that is,

$$G = 0 \quad \text{on} \quad S \quad \text{(Dirichlet)} \tag{1-38}$$

$$\partial G/\partial n = 0 \quad \text{on} \quad S \quad \text{(Neumann)} \tag{1-39}$$

$$G + h(\partial G/\partial n) = 0 \quad \text{on} \quad S \quad \text{(mixed)} \tag{1-40}$$

where h is a parameter that keeps the dimensions consistent.[†] The condition appropriate to this work will be the Dirichlet condition since the electric fields, which have zero tangential fields at the boundary, are being solved for. If the Dirichlet condition is applied, then Eq. (1-37) reduces to

$$u(r') = \iint_{-\infty}^{\infty} u(x, y, z)(\partial G/\partial n)\, dx\, dy \bigg|_{z=0} \tag{1-41}$$

where the flat plane is aligned appropriately with the xy plane.[‡]

To complete the solution for $u(r')$, a Green's function needs to be found which will have the properties that correspond to the physical constraints of the particular medium. First, a plausibility argument for the form G can be given. Consider the Green's function equation over a region of space that does not include the source point r' as shown in Fig. 1-4. Further, consider the space to have angular symmetry. The equation then reduces to

$$(\nabla^2 + k^2)G = \frac{1}{r^2}\frac{\partial}{\partial r}\left(r^2 \frac{\partial G}{\partial r}\right) + k^2 G = 0, \qquad r \neq r' \tag{1-42}$$

One cannot generally recognize the solution to this equation; thus, it is

[†] In electromagnetic problems, h has the dimension of ohms.

[‡] The plus sign arises because ∇G is outward in direction and on the xy plane $\hat{n}$ is in plus z direction or inward to the surface.

desirable to convert it to an equation that is recognizable, namely,

$$(\partial^2/\partial x^2 + k^2)G = 0 \tag{1-43}$$

This transformation can be accomplished by defining a function† $f(r)$ as [15]

$$f(r)/r = G(r) \tag{1-44}$$

Substitution of Eq. (1-44) into Eq. (1-42) gives

$$d^2f/dr^2 + k^2f = 0, \qquad r \neq r' \tag{1-45}$$

The solution to this equation is given by Eq. (1-46), leaving only the problem of finding the appropriate constants C_1 and C_2

$$f(r) = C_1e^{ikr} + C_2e^{-ikr} \tag{1-46}$$

The solution for $G(r)$ can be written in terms of C_1 and C_2

$$G(r) = C_1(e^{ikr}/r) + C_2(e^{-ikr}/r) \tag{1-47}$$

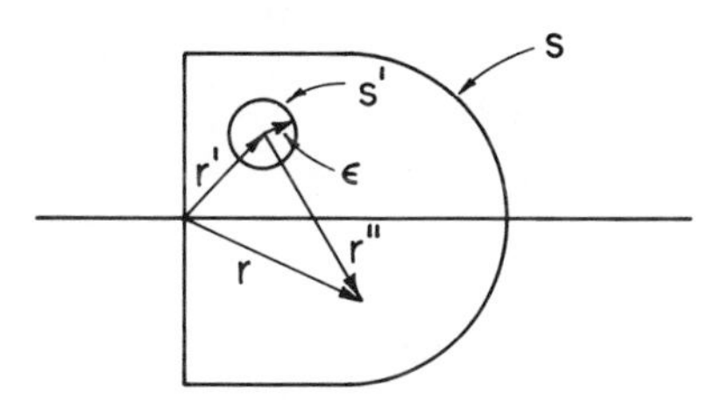

Fig. 1-4. Right semi-infinite half space with a volume enclosed by S' excluded to find the constant $1/4\pi$. The radius of the excluded volume is ε.

Since it was assumed a priori that the time dependence varies like $e^{-j\omega t}$, the first term corresponds to an outward traveling wave in a physical radiating system. The second term corresponds to an inward traveling wave that might have arisen from a reflection. Because only outward traveling waves are being considered, the constant C_2 can be set equal to zero.

To find the constant C_1, consider the region of space that encloses the point r' shown in Fig. 1-4. If a volume integration of the inhomogeneous form of Eq. (1-42) is performed about this point with a radius of ε and the

† Transformations of this type are common in spherical coordinates.

Laplacian is converted by using the divergence theorem, then the following is obtained:

$$\lim_{\varepsilon\to 0}\left[\iint_{S'} \nabla G\cdot \hat{n}\, dS + k^2\int_{V_\varepsilon} G\, dV = -\int_{V_\varepsilon} \delta(r-r')\, dV = -1\right] \quad (1\text{-}48)$$

In the limit of zero volume, the second term on the left goes to zero, leaving†

$$\lim_{\varepsilon\to 0}\left[C_1\int_0^{2\pi}\int_0^{\pi} d\theta\, d\phi r''^2 \sin\theta \frac{\partial}{\partial r''}\left(\frac{e^{jkr''}}{r''}\right)\right]\Bigg|_{r''=\varepsilon} = -1 \quad (1\text{-}49)$$

where the reduced form of Eq. (1-47) has been substituted. The angular integrations are straightforward, leading to 4π, reducing Eq. (1-49) to‡

$$\lim_{\varepsilon\to 0}\left[4\pi C_1 r''^2 \frac{\partial}{\partial r''}\left(\frac{e^{jkr''}}{r''}\right)\right]\Bigg|_{r''=\varepsilon} = -1 \quad (1\text{-}50)$$

This equation reduced to

$$\lim_{\varepsilon\to 0}\,[4\pi C_1 e^{jkr''}(jkr''-1)]\,|_{r''=\varepsilon} = -1 \quad (1\text{-}51)$$

which under the limit of $r'' = \varepsilon \to 0$ gives

$$C_1 = 1/4\pi \quad (1\text{-}52)$$

Thus, an expression for $G(r)$ that corresponds to an infinite half-space radiation problem is obtained:

$$G(r'') = e^{jkr''}/4\pi r'' \quad (1\text{-}53)$$

Physically, a wave of this functional form is spherical in nature due to the $1/r''$ dependence and has a linear phase change represented by the exponent term.

It is appropriate at this point to expand upon some of the properties of such Green's functions. First, note that in general the coordinate system is to be changed back to one including r' as shown in Fig. 1-4. The basic properties are:

1. The Green's function is symmetric with respect to r and r'. Thus, $G(r, r') = G(r', r)$, which may be obvious from the differential equation but can be further thought of as a result of the reciprocity theorem.

† It is important to note that there is an implicit change of coordinates to r'' centered about the point r'. This simplifies the mathematics for the brief derivation, but one must convert back following the convention $r'' \to r - r'$.

‡ It is valid to use Eq. (1-47) for $G(r'')$ since on the surface of the elemental volume this $G(r)$ applies.

2. The first derivative of the Green's function has a jump at $r = r'$. From the differential equation, Eq. (1-32), note that the second derivative of G has the behavior of a delta function; therefore, the first derivative has a discontinuity like a step.

3. The Green's function is a continuous function of r when r' is fixed, even when $r = r'$.

These properties are intuitively obvious from work in pulse circuits and will not be explored further at this point.†

To use this Green's function in solving Eq. (1-41), refer to the coordinate system shown in Fig. 1-5 and write the function in terms of r_s and r_o' as‡

$$G(r_o, r_s) = \frac{e^{jk|r_o' - r_s|}}{4\pi \mid r_o' - r_s \mid} \tag{1-54}$$

Further, the particular function for this problem must also satisfy the Dirichlet boundary condition at the surface ($G = 0$ on S). This condition can be established by using an image source on the other side of S at a point r_s''. This is a symmetric point to r_s, and whenever r_s is moved closer to the surface S, so is r_s''; that is, r_s'' is a function of r_s. In the limit r_s goes to zero and the generalized origin moves to the coordinate center of S. Thus $G(r_s, r_o')$ becomes§

$$G(r, r') = \frac{e^{jk|r_o' - r_s|}}{4\pi \mid r_o' - r_s \mid} - \frac{e^{jk|r_o' - r_s''|}}{4\pi \mid r_o' - r_s'' \mid} = G_1 - G_2 \tag{1-55}$$

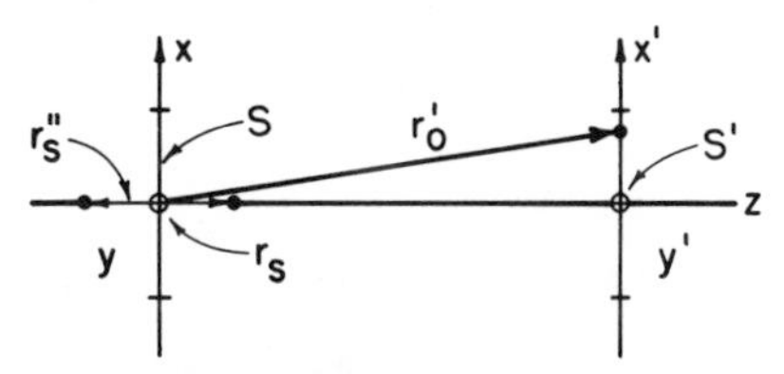

Fig. 1-5. General geometry and image source location in order to satisfy the Dirichlet boundary condition ($G = 0$) over the transmitting aperture.

To find the normal derivative of G used in Eq. (1-41), take the derivative as follows:

$$\frac{\partial G}{\partial n} = \frac{\partial G}{\partial z} = \frac{\partial G_1}{\partial z} - \frac{\partial G_2}{\partial z} = \frac{\partial G_1}{\partial R'} \frac{\partial R'}{\partial z} - \frac{\partial G_2}{\partial R''} \frac{\partial R''}{\partial z} \tag{1-56}$$

† The reader is referred to other texts for further details [3, 11].

‡ Remember that r_o' applies to the observation aperture and r_s applies to the excitation aperture.

§ See the problem set for an alternate approach.

where R is the distance between the observation point and each source point, that is,

$$R'^2 = (x' - x_s)^2 + (y' - y_s)^2 + (z' - z_s)^2 \tag{1-57a}$$

$$R''^2 = (x' - x_s'')^2 + (y' - y_s'')^2 + (z' + z_s'')^2 \tag{1-57b}$$

Now

$$\frac{\partial G_1}{\partial R'} (=) \frac{\partial G_2}{\partial R''} \tag{1-58}$$

in form, giving

$$\frac{\partial G}{\partial z} = - \frac{e^{jkR'}(jkR' - 1)}{4\pi R'^3} (z'-z) - \frac{e^{jkR''}(jkR'' - 1)}{4\pi (R'')^3} (z'+z) \tag{1-59}$$

Remember that in this case $z = 0$, because both sources go to the surface. This considerably simplifies the result. Thus, Eq. (1-59) becomes

$$\left.\frac{\partial G}{\partial z}\right|_{z=0} = - \frac{2z'}{4\pi R^3} e^{jkR}(jkR - 1) \tag{1-60}$$

where

$$R = \sqrt{(x' - x)^2 + (y' - y)^2 + z'^2} \tag{1-61}$$

Equation (1-41) now becomes

$$u(x', y', z') = \frac{1}{2\pi} \iint_{-\infty}^{\infty} u(x, y, 0) \left(\frac{z'}{R}\right)\left(\frac{1}{R} - jk\right) \frac{e^{jkR}}{R} \, dx \, dy \tag{1-62}$$

Examination of the terms reveals a simpler representation. First, the z'/R term is replaced by†

$$z'/R = \cos(z', R) \tag{1-63}$$

which, if it is assumed that $R \gg \lambda$, or $1/R \ll 1/\lambda$, then

$$u(x', y', z') = - \frac{j}{\lambda} \iint_{-\infty}^{\infty} u(x, y, 0) \frac{e^{jkR}}{R} \cos(z', R) \, dx \, dy \tag{1-64}$$

One additional approximation that is commonly made is that $z', R \gg x', y', x, y$, reducing $\cos(z', R)$ to unity, which leaves

$$u(x', y', z') = - \frac{j}{\lambda} \iint_{-\infty}^{\infty} u(x, y, 0) \frac{e^{jkR}}{R} \, dx \, dy \tag{1-65}$$

† This use of the cosine could have been seen directly from the directional derivative, i.e., $\partial G/\partial n = (\partial G/\partial R) \cos(n, R)$.

Note the similarity of this result with that given by Eq. (1-31). It has been shown that the propagator $K(\ )$ is indeed in the approximate sense, a spherical wave radiator and that the constants out front are like those assumed by Fresnel and Huygens, $(-j/\lambda)$. This result, commonly referred to as the Rayleigh–Sommerfeld integral [11, 15], will form the basis of the later work in this book.†

A similar, less rigorous approach was developed by Kirchhoff, and it retains both terms in Eq. (1-37) and uses some special ad hoc boundary conditions on u and $\partial u/\partial n$ to obtain a similar expression. See Goodman's book [13] for this discussion.

Problems

1. Expand $\nabla\times\nabla\times\mathbf{A}$ in the cylindrical and spherical coordinate systems and show that $\nabla_{\mathrm{v}}^2 \neq \nabla_{\mathrm{s}}^2$.

2. Derive the wave equation for **H**.

3. Derive the wave equations for **E** and **H** in nonsource-free regions of space. Discuss what one obtains if ε is a function of position.

4. Demonstrate that $\nabla^2(\log u) + (\nabla \log u)^2 = \nabla^2 u/u$.

5. Demonstrate the equation

$$\nabla^2(\log A) + (\nabla \log A)^2 - (\nabla S)^2 + k_0^2 n^2 + i(\nabla^2 S + 2\nabla \log A \cdot \nabla S) = 0$$

6. Using $\nabla \cdot \phi\mathbf{A} = \phi\nabla \cdot \mathbf{A} + \mathbf{A} \cdot \nabla\phi$ show that

$$\nabla \cdot \phi\nabla\psi = \phi\nabla^2\psi + \nabla\psi \cdot \nabla\phi$$

and further that

$$\nabla \cdot (\phi\nabla\psi - \psi\nabla\phi) = \phi\nabla^2\psi - \psi\nabla^2\phi$$

7. Using the divergence theorem, prove Green's theorem, namely, that

$$\iiint_V (u\,\nabla^2 v - v\,\nabla^2 u)\,dV = \oiint_S (u\,\nabla v - v\,\nabla u) \cdot \hat{n}\,dS$$

where u and v are scalars and V indicates a volume.

† Note that as R gets very large, such as in the Fraunhofer zone, that the amplitude dependence on R, i.e., $1/R$, can be taken outside and $u(x', y', z')$ becomes similar to a Fourier transform expression for $u(x, y)$ when quadratic phase terms are neglected. These implications will be explored further in the next chapter.

8. If the generalized curvilinear del operator and divergence operator are defined as

$$\nabla = \frac{\hat{i}_1}{h_1}\frac{\partial}{\partial u_1} + \frac{\hat{i}_2}{h_2}\frac{\partial}{\partial u_2} + \frac{\hat{i}_3}{h_3}\frac{\partial}{\partial u_3}$$

and

$$\nabla \cdot \mathbf{F} = \frac{1}{h_1 h_2 h_3}\left[\frac{\partial}{\partial u_1}(h_2 h_3 f_1) + \frac{\partial}{\partial u_2}(h_3 h_1 f_2) + \frac{\partial}{\partial u_3}(h_1 h_2 f_3)\right]$$

Derive the general form of the gradient of ϕ, $\nabla\phi$, and the Laplacian of ϕ, $\nabla^2\phi$, for the cylindrical and spherical coordinate systems.

9. Show by direct substitution that Eq. (1-45) results when Eq. (1-44) is used in Eq. (1-42).

10. Show how Eq. (1-59) follows from Eqs. (1-55) through (1-58).

11. Using the Neumann condition on G, show that

$$G(r, r') = \frac{1}{4\pi}\left[\frac{e^{jk|r_o' - r_s|}}{|r_o' - r_s|} + \frac{e^{jk|r_o' - r_s''|}}{|r_o' - r_s''|}\right]$$

is the appropriate function to make $\partial G/\partial n$ on the surface equal to zero.

Chapter II

APPLICATIONS AND APPROXIMATIONS FOR RADIATION FIELDS

Introduction

In Chapter I an expression for the field $u(x', y', z')$ at a point in space due to an exciting field $u(x, y, z)$ at some removed point was derived:

$$u(x', y', z') = -\frac{j}{\lambda} \iint_{-\infty}^{\infty} u(x, y, z) \frac{e^{jkR}}{R} \, dx \, dy \bigg|_{z=0} \tag{2-1}$$

where

$$R = [(x' - x)^2 + (y' - y)^2 + z'^2]^{1/2} \tag{2-2}$$

This equation for large R reduces to a form similar to the Fourier transform [15] when quadratic phase terms are neglected. This is not quite correct in an explicit sense since the exponent argument is not in the proper form.

This assertion will be examined in detail for various regions of space. These regions will become associated with various approximations. The first approximation is connected with the idea that sources and observation points will be separated by a distance which is larger than any linear aperture dimension; or specifically z' is very much larger than x, x', y, or y'. With this constraint R can be expanded in a binomial expansion giving

$$\begin{aligned} R &= z'\left[1 + \frac{(x' - x)^2}{z'^2} + \frac{(y' - y)^2}{z'^2}\right]^{1/2} \\ &\cong z' + \frac{(x' - x)^2}{2z'} + \frac{(y' - y)^2}{2z'} \end{aligned} \tag{2-3}$$

Expanding Eq. (2-3) and collecting like terms yields

$$R = z' - \frac{x'x + y'y}{z'} + \frac{x'^2 + y'^2}{2z'} + \frac{x^2 + y^2}{2z'} \tag{2-4}$$

Equation (2-2) can now be replaced by Eq. (2-4), which is z' plus transverse quadratic terms. Since the integration is over x and y, some of the quadratic terms can be taken outside the integral. Before doing this, note that the $1/R$ amplitude term can be approximated by $1/z'$ when the assumptions leading to Eq. (2-3) are used. This cannot be done in the exponent phase term since any resulting errors in neglecting the transverse terms are enhanced by the large multiplying factor k. Therefore, the following expression for $u(x', y', z')$ results

$$\begin{aligned} u(x', y', z') = {} & \frac{e^{jkz'} \exp[j(k/2z')(x'^2 + y'^2)]}{j\lambda z'} \\ & \times \iint_{-\infty}^{\infty} u(x, y, 0) \exp[j(k/2z')(x^2 + y^2)] \\ & \times \exp[-j(k/z')(xx' + yy')]\, dx\, dy \end{aligned} \tag{2-5}$$

This approximation is referred to as the Fresnel approximation. It basically is the replacement of the square root by the first two terms of the binomial expansion. Except for the quadratic phase terms, and complex constants, the field at (x', y', z') is written as a Fourier transform of the field at $u(x, y, 0)$ where the spatial frequency variables are

$$u = kx'/z' \qquad \text{and} \qquad v = ky'/z' \tag{2-6}$$

Even in this representation the integrand is not exactly in the form of a Fourier transform [13]. The integrand still contains a quadratic phase term over the exciting aperture variables. If this term is neglected, then the integral is in the explicit form of a Fourier transform. This approximation is referred to as the Fraunhofer approximation and is written as

$$\begin{aligned} u(x', y', z') = {} & \frac{e^{jkz'} \exp[j(k/2z')(x'^2 + y'^2)]}{j\lambda z'} \\ & \times \iint_{-\infty}^{\infty} u(x, y, 0) \exp[-j(k/z')(xx' + yy')]\, dx\, dy \end{aligned} \tag{2-7}$$

where it has been assumed that

$$\exp[j(k/2z')(x^2 + y^2)] \simeq 1 \tag{2-8}$$

Equation (2-8) provides a format for examining a defining region between the Fresnel and Fraunhofer approximation. Note that e^{jx} is of order unity[†] for $x \leq 0.5$ radians. Applying this to Eq. (2-8) yields

$$(k/2z')(x^2 + y^2) \simeq 0.5 \tag{2-9}$$

If $\mathscr{R}$ is a nominal linear measure of the aperture ($\mathscr{R}^2 = x^2 + y^2$), then

$$\pi\mathscr{R}^2/\lambda z_f' \simeq 0.5 \tag{2-10}$$

Recognizing that this is a loose statement regarding the boundary, it is probably sufficient to say that the left-hand side of Eq. (2-10) is of order unity[‡]

$$\pi\mathscr{R}^2/\lambda z_f' = 1 \tag{2-11}$$

Often the statement is written as

$$\mathscr{R}^2/\lambda z_f' \simeq 1 \tag{2-12}$$

which arises from a classical optics approach. Throughout this book Eq. (2-11) will be used as the defining relation for z_f, the Fresnel–Fraunhofer boundary.

To illustrate what this distance is like, consider the case of visible light emanating from a 1-mm aperture, then

$$z_f = \frac{\pi\mathscr{R}^2}{\lambda} = \frac{(\pi)(0.001 \text{ m})^2}{0.6\times 10^{-6} \text{ m}} \simeq 5 \text{ m} \tag{2-13}$$

If $\mathscr{R}$ is 1 cm, then

$$z_f = 500 \text{ m} \tag{2-14}$$

These examples illustrate that small changes in aperture can change z_f significantly.

Illustration of Fraunhofer Fields

To illustrate the results, consider a rectangular aperture of dimension $2a\times 2b$ having a constant field of amplitude A in the aperture, as shown in

[†] That is, for an error of less than 6 percent.

[‡] This result is consistent with a beam-wave derivation [10] where the beam diameter W_0 decreases by $\sqrt{2}$, or physically the half power point.

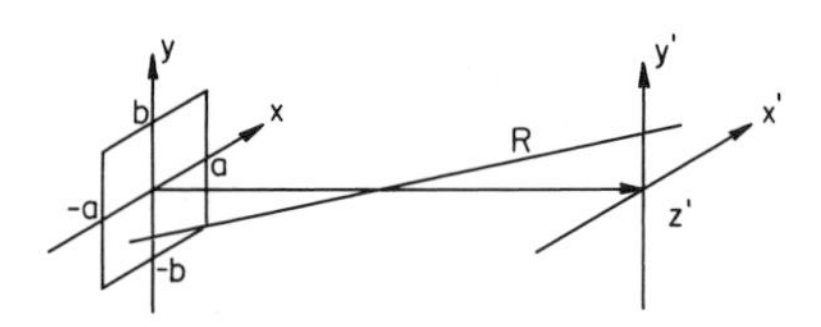

Fig. 2-1. Geometry for a rectangular aperture ($2a \times 2b$) that radiates a distance R to a plane (x', y').

Fig. 2-1. Using Eq. (2-7) in the Fraunhofer region yields

$$u(x', y', z') = \left[\frac{Ae^{jkz'} \exp[j(k/2z')(x'^2 + y'^2)]}{j\lambda z'}\right]$$

$$\times \int_{-b}^{b} \int_{-a}^{a} \exp[-j(k/z')(xx' + yy')]\, dx\, dy \tag{2-15a}$$

$$= [\ ]\left(\frac{\exp[j(kx'a/z')] - \exp[-j(kx'a/z')]}{j(kx'/z')}\right)$$

$$\times \left(\frac{\exp[j(ky'b,z')] - \exp[-j(ky'b/z')]}{j(ky'/z')}\right) \tag{2-15b}$$

$$= 4ab[\ ]\left(\frac{\sin(kx'a/z')}{kxa'/z'}\right)\left(\frac{\sin(ky'b/z')}{ky'b/z'}\right) \tag{2-15c}$$

Now the terms in the large parentheses are the familiar $(\sin \xi)/\xi$ type terms that give the familiar plots shown in Fig. 2-2. These effects are primarily determined by the aperture dimensions.

If direct detection of this optical energy distribution is done, then the intensity distribution results

$$I(x', y', z') = u(x', y', z')u^*(x', y', z') \tag{2-16}$$

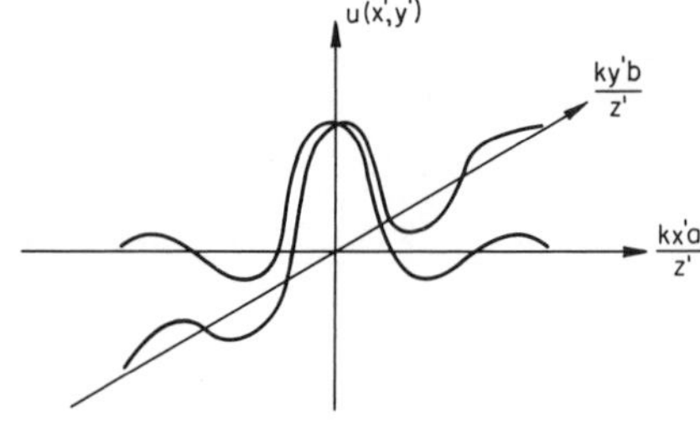

Fig. 2-2. Two-dimensional amplitude distribution of Fraunhofer field for a rectangular aperture of dimensions $2a \times 2b$.

Substituting Eq. (2-15) into (2-16) yields

$$I(x', y', z') = \frac{(4ab)^2 A^2}{\lambda^2 z'^2} \left(\frac{\sin(kx'a/z')}{kx'a/z'} \right)^2 \left(\frac{\sin(ky'b/z')}{ky'b/z'} \right)^2 \tag{2-17}$$

which can be rewritten as

$$I(x', y', z') = \frac{(4ab)^2 A^2}{\lambda^2 z'^2} I_n(x', y', z') \tag{2-18}$$

where $I_n(x', y', z')$ is a normalized intensity shown in Fig. 2-3. Note that all the complex exponential terms outside the sinc functions dropped out due to the conjugation operation. Physically this is because the intensity has to be a real function if it is an observable one.

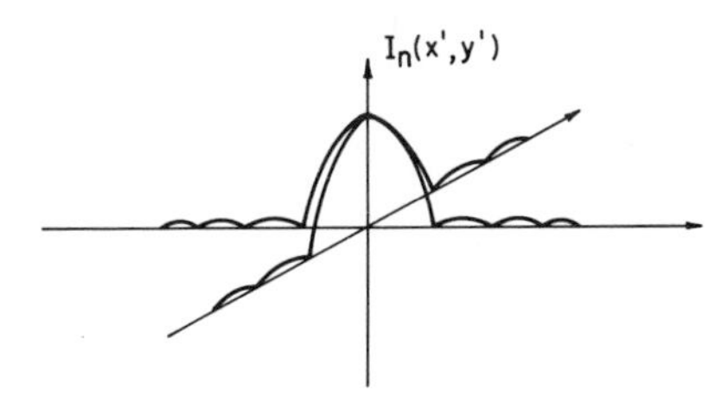

Fig. 2-3. Intensity distribution pattern for the Fraunhofer field of a rectangular aperture shown in Fig. 2-2 and Fig. 2-1.

Circular Aperture

It is interesting to consider the circular-aperture case because of some of the mathematical operations that may be unfamiliar to some readers. Here the aperture field is defined as

$$u(r) = \begin{cases} A & r < a \\ 0 & r \geq a \end{cases} \tag{2-19}$$

where $r = [x^2 + y^2]^{1/2}$. Recognizing the angular symmetry, Eq. (2-7) can be transformed using the following definitions:

$$\begin{aligned} x &= r \cos \theta, \qquad x' = r' \cos \theta', \qquad dx\, dy \to r\, dr\, d\theta \\ y &= r \sin \theta, \qquad y' = r' \sin \theta', \end{aligned} \tag{2-20}$$

and the identity

$$\sin \theta \sin \theta' + \cos \theta \cos \theta' = \cos(\theta - \theta') \tag{2-21}$$

Because of the argument of $\cos(\theta - \theta')$, the limits[†] of the θ integration are chosen to be θ' to $2\pi + \theta'$. The alignment of the two coordinate frames is arbitrary and chosen for convenience. Now $u(r)$ is

$$u(r') = \frac{Ae^{jkz'} \exp[j(kr'^2/2z')]}{j\lambda z'} \times \int_0^a \int_{\theta'}^{2\pi+\theta'} \exp\left[-j \frac{2\pi r r' \cos(\theta - \theta')}{\lambda z'}\right] r\, dr\, d\theta \tag{2-22}$$

where the θ integration will be performed first. Using the change of variable

$$\theta - \theta' = u \tag{2-23}$$

the θ integration becomes

$$\int_0^{2\pi} \exp\left[-j \frac{2rr'}{\lambda z'} \cos u\right] du \tag{2-24}$$

which if the following integral definition of a Bessel function [16] is used

$$J_n(z) = \frac{i^{-n}}{2\pi} \int_0^{2\pi} e^{-iz\cos\phi} e^{in\phi}\, d\phi \tag{2-25}$$

Eq. (2-24) becomes

$$2\pi J_0(2\pi r r'/\lambda z') \tag{2-26}$$

The field $u(r')$ is now reduced to an integral over r,

$$u(r') = \frac{A 2\pi e^{jkz'} \exp[j(kr'^2/2z')]}{j\lambda z'} \int_0^a r J_0\left(\frac{2\pi r r'}{\lambda z'}\right) dr \tag{2-27}$$

Using the identity [16]

$$\int x Z_0(x)\, dx = x Z_1(x) \tag{2-28}$$

and a change of variables, $u(r')$ becomes

$$u(r') = \frac{Aae^{jkz'} \exp[j(kr'^2/2z')]}{jr'} J_1\left(\frac{2\pi r' a}{\lambda z'}\right) \tag{2-29}$$

Since the function $J_1(x)/x$ is a function similar to $\sin x/x$, Eq. (2-29) can be rewritten in the following form:

$$u(r') = \frac{2AB}{j\lambda z'} e^{jkz'} \exp\left(j \frac{kr'^2}{2z'}\right) \frac{J_1(\xi r')}{\xi r'} \tag{2-30}$$

[†] This choice is valid since our functions of θ are periodic in 2π and the orientation of θ with respect to θ' is arbitrary.

where B is the area of the aperture πa^2, and ξ equals ka/z', the radial spatial frequency. A plot of $J_1(\xi r')/\xi r'$ is shown in Fig. 2-4. From this result the origin of the diameter of the classical Airy disk can be seen. To do so, the r' characterizing the first zero or the border of the central disk r_d' will be solved for. Setting the argument $2ar'/\lambda z'$ of $J_1(\ )$ equal to 1.22, the Airy disk diameter $2r_d'$ is found to be equal to $1.22\lambda z'/a$. Notice now that the functional form of the field is the same in this coordinate system as in the rectangular system. It involves the area (πa^2 or ab) in the numerator, $\lambda z'$ in the denominator, a linear phase term in kz', and a quadratic phase term in the transverse coordinates. Similarly $J_1(x)/x$ is a function very much like $\sin(x)/x$, where the nulls are shifted slightly. The reader can verify for himself that the intensity for the rectangular case would be similar.

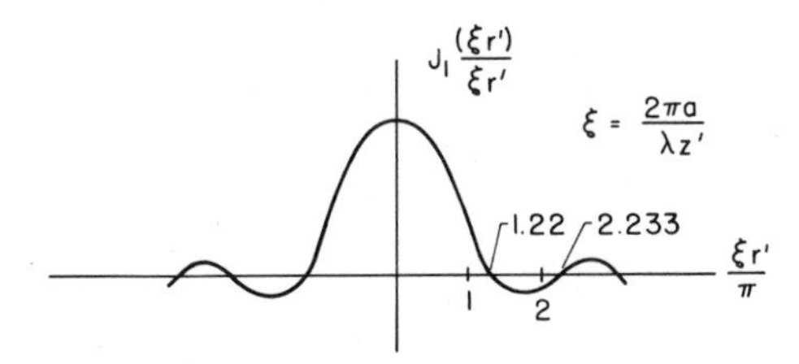

Fig. 2-4. One-dimensional amplitude in the radial direction for a circular aperture far-field pattern.

Multiple Apertures—Array and Aperture Factors

So far only the radiated fields associated with single apertures have been discussed; consideration of how several apertures would work together has been ignored. This becomes particularly important when the apertures are coherently illuminated or illuminated in phase, e.g., by the same plane wave. Referring to Fig. 2-5, assume that each aperture is illuminated by the plane wave of amplitude A. Using Eq. (2-7) and following Eq. (2-15), we obtain the following:

$$u(x', y', z') = \left[\frac{Ae^{jkz'}\exp[j(k/2z')(x'^2 + y'^2)]}{j\lambda z'}\right] \times \left[\int_{-d/2-b}^{-d/2+b}\int_{-a}^{a} \exp\left[-j\frac{k}{z'}(xx' + yy')\,dx\,dy\right] + \int_{d/2-b}^{d/2+b}\int_{-a}^{a} \exp\left[-j\frac{k}{z'}(xx' + yy')\,dx\,dy\right]\right] \quad (2\text{-}31)$$

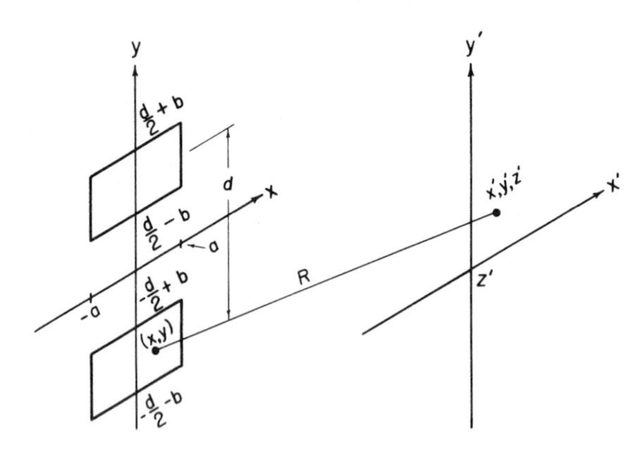

Fig. 2-5. Geometry for multiple-aperture analysis. Two rectangular apertures ($2a \times 2b$) are spaced a distance d apart.

This equation is very much like Eq. (2-15), except that now there are two integrals inside the second square bracket. Notice further that the x integration is the same in each case and can be factored out. It is also the same as the x integration in Eq. (2-15). If appropriate terms are factored and collected, Eq. (2-31) becomes

$$u(x', y', z') = [\ ]\left[\int_{-a}^{a} \exp\left(-j\frac{k}{z'} xx'\, dx\right)\right]\left[\int_{-d/2-b}^{-d/2+b} \exp\left(-j\frac{k}{z'} yy'\, dy\right) + \int_{d/2-b}^{d/2+b} \exp\left(-j\frac{k}{z'} yy'\right) dy\right] \tag{2-32}$$

The third square bracket term can also be reduced to a more recognizable form by defining new variables in the first and second terms as

$$y_1 = y + d/2 \rightarrow dy_1 = dy \tag{2-33a}$$

and

$$y_2 = y - d/2 \rightarrow dy_2 = dy \tag{2-33b}$$

which reduces the third square bracket term of Eq. (2-32) to

$$\begin{aligned} u(x', y', z') = [\ ]&\left[\int_{-a}^{a} \exp\left(-j\frac{k}{z'} xx'\, dx\right)\right] \\ &\times \left[\exp\left(j\frac{kd}{2z'} y'\right)\int_{-b}^{b} \exp\left(-j\frac{k}{z'} y'y_1\, dy_1\right)\right. \\ &\left. + \exp\left(-j\frac{kd}{2z'} y'\right)\int_{-b}^{b} \exp\left(-j\frac{k}{z'} y'y_2\, dy_2\right)\right] \end{aligned} \tag{2-34}$$

Now note that the integrals inside the third bracket are identical. If terms are factored and collected together, Eq. (2-34) can be rewritten as

$$u(x', y', z') = [\ ]\left[\int_{-b}^{b}\int_{-a}^{a} \exp\left[-j\frac{k}{z'}(xx' + yy')\,dx\,dy\right]\right]$$
$$\times \left\{\exp\left(j\frac{k\,dy'}{2z'}\right) + \exp\left(-j\frac{k\,dy'}{2z'}\right)\right\} \tag{2-35}$$

where the last brace term can be written, using the Euler relations, as

$$\left\{\exp\left(j\frac{k\,dy'}{2z'}\right) + \exp\left(-j\frac{k\,dy'}{2z'}\right)\right\} = 2\cos\left(\frac{k\,dy'}{2z'}\right) \tag{2-36}$$

Thus Eq. (2-35) is identical to Eq. (2-15) except for the added multiplying factor $2\cos(k\,dy'/2z')$, which is called the array factor. Using Eq. (2-15) and Eq. (2-36), we can write the two aperture fields as

$$u(x', y', z') = 4ab[\ ]\left(\frac{\sin(kx'a/z')}{kx'a/z'}\right)\left(\frac{\sin(ky'b/z')}{ky'b/z'}\right)\left(2\cos\left(\frac{k\,dy'}{2z'}\right)\right) \tag{2-37}$$

which gives a field distribution as shown in Fig. 2-6. For illustrative purposes it has been assumed that $d = 4b$. Notice the similarity to Fig. 2-2.

The effect of the array factor is to provide an envelope function in the y' direction. It reduces the side-lobe amplitudes of the sinc function in the

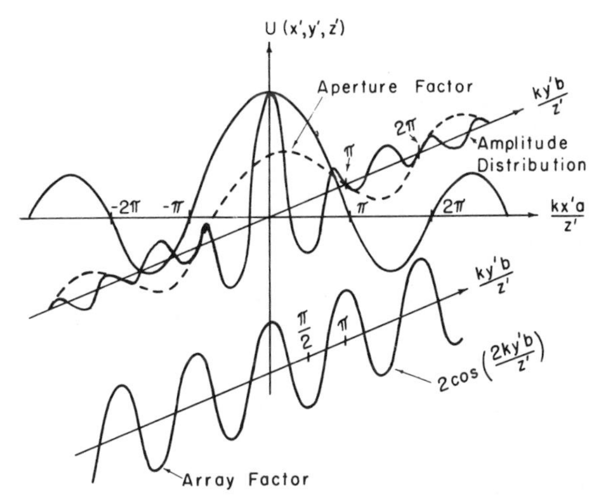

Fig. 2-6. Two-dimensional amplitude distribution of the Fraunhofer field of two symmetric rectangular apertures. It has been assumed that $d = 4b$ for purposes of illustration. The dotted line is the aperture factor alone. The array factor is shown in a plot just below the amplitude distribution.

y' direction and adds appropriate zeros corresponding to the zeros of the array factor.

The two sinc functions over x' and y', taken together, are defined as the aperture factor since they are related specifically to the aperture characteristics. In fact, even if additional coherent apertures are added, this aperture-factor term will remain the same. Only the array factor will change. It is easy to see that plotting more than a couple apertures becomes a difficult problem. However, the properties of the array factors can generally be written down by inspection since the mathematical procedure is the same no matter how many are present. What one generally ends up with is a cosine series if things are symmetric or a sine series if odd symmetries are involved.

These concepts are directly applicable to antenna theory and help to explain side-lobe behavior. In particular, it is possible to see how side-lobe level and main-lobe width can be reduced by added array elements.

If N apertures are considered, the case of a classic grating is obtained [11]. The array-factor series can then be written in closed form,[†] giving another simple geometry problem with very sharp lobe structure for easy identification of diffraction orders. In the problem set additional examples will be considered.

Phase Effects in Apertures—Wedge

As another example of Fraunhofer patterns, consider the case of an aperture which has a special phase retardation plate of the form

$$t(x, y) = \begin{cases} \exp\{j[\beta(x+a)+\gamma(y+b)]\}, & -a < x < a, \ -b < y < b \\ 0 & \text{otherwise} \end{cases} \tag{2-38}$$

Physically, this represents a linearly delayed phase across the aperture as the wave traverses the aperture. The field is then

$$u(x', y', z') = \frac{e^{jkz'} \exp[j(k/2z')(x'^2 + y'^2)]}{j\lambda z'} \times \int_{-a}^{a} \int_{-b}^{b} \exp\{j[\beta(x + a) + \gamma(y + b)]\} \times \exp\left[-j\frac{k}{z'}(xx' + yy')\right] dx\, dy \tag{2-39}$$

[†] An important formula that is used in obtaining the closed-form representations is [17]

$$1 + 2\cos 2t + 2\cos 4t + \cdots + 2\cos 2nt = \frac{\sin(2n+1)t}{\sin t}$$

which can be rewritten as

$$u(x', y', z') = \frac{\exp[j(kz' + \beta a + \gamma b)] \exp[j(k/2z')(x'^2 + y'^2)]}{j\lambda z'} \times \int_{-a}^{a} \int_{-b}^{b} \exp\left\{-j\left[\left(\frac{kx'}{z'} - \beta\right)x + \left(\frac{ky'}{z'} - \gamma\right)y\right]\right\} dx\, dy \tag{2-40}$$

This integral occurred before in the development leading to Eq. (2-15). If a simple coordinate shift is considered, the result can be written by inspection, giving

$$u(x', y', z') = \frac{4abe^{jkz'} \exp[j(k/2z')(x'^2 + y'^2)]}{j\lambda z'} \exp[j(\beta a + \gamma b)] \times \frac{\sin[(kx'/z' - \beta)a] \sin[(ky'/z' - \gamma)b]}{ab(kx'/z' - \beta)(ky'/z' - \gamma)} \tag{2-41}$$

This result is just like the result in Eq. (2-15) for the plain square aperture, except for the fixed phase-delay term $\exp[j(\beta a + \gamma b)]$, and the shift in peak of the sinc functions to a new point defined by

$$(x', y')_{\text{peak}} = (z'\beta/k, z'\gamma/k) \tag{2-42}$$

This simply amounts to a shift in the diffraction pattern or a steering of the wave in a linear fashion. Thus a wedge gives some delay and a linear shift in the position of the peak or major lobe.

Fresnel Approximation

As an example of a Fresnel approximation, consider an aperture that has a phase delay of the following form:

$$t(x, y) = \begin{cases} \exp[-j(\beta x^2 + \gamma y^2)], & -a < x < a, \quad -b < y < b \\ 0 & \text{otherwise} \end{cases} \tag{2-43}$$

The field that results is

$$u(x', y', z') = \frac{e^{jkz'} \exp[j(k/2z')(x'^2 + y'^2)]}{j\lambda z'} \times \int_{-a}^{a} \int_{-b}^{b} \exp[-j(\beta x^2 + \gamma y^2)] \exp\left[j\frac{k}{2z'}(x^2 + y^2)\right] \times \exp\left[-j\frac{k}{z'}(xx' + yy')\right] dx\, dy \tag{2-44}$$

which upon rearranging becomes

$$u(x', y', z') = \frac{e^{jkz'} \exp[j(k/2z')(x'^2 + y'^2)]}{j\lambda z'}$$
$$\times \int_{-a}^{a} \int_{-b}^{b} \exp\left\{j\left[\left(\frac{k}{2z'} - \beta\right)x^2 + \left(\frac{k}{2z'} - \gamma\right)y^2\right]\right\}$$
$$\times \exp\left[-j\frac{k}{z'}(x'x + y'y)\right] dx\, dy \qquad (2\text{-}45)$$

Note now that if the special case of z' equals $k/2\beta$ and β equals γ is considered, this integral reduces to the Fraunhofer pattern of a plain rectangular aperture. Furthermore, the field is just the Fourier transform of the exciting aperture transmittance function. This will be shown later to be the case of an angularly symmetric thin lens, where the observation point is in the focal plane of the lens.

To consider a case other than this special one, the following change in variables is used:

$$\xi = k/2z' - \beta \qquad (2\text{-}46)$$

$$\eta = k/2z' - \gamma \qquad (2\text{-}47)$$

Appropriately factoring the integrand term as

$$\int_{-a}^{a} \exp\left[j\xi\left(x^2 - \frac{kx'}{z'\xi}x\right)\right] dx \int_{-b}^{b} \exp\left[j\eta\left(y^2 - \frac{ky'}{z'\eta}y\right)\right] dy \qquad (2\text{-}48)$$

and defining new variables g and h as

$$g^2 = \frac{2}{\pi}\,\xi\left(x - \frac{kx'}{2z'\xi}\right)^2 \qquad (2\text{-}49)$$

$$h^2 = \frac{2}{\pi}\,\eta\left(y - \frac{ky'}{2z'\eta}\right)^2 \qquad (2\text{-}50)$$

yields

$$\exp\left[-j\frac{k^2}{4z'^2}\left(\frac{x'^2}{\xi} + \frac{y'^2}{\eta}\right)\right]\left(\frac{\pi}{2\xi}\right)^{1/2} \int_{-(2\xi/\pi)^{1/2}(a+kx'/2z'\xi)}^{(2\xi/\pi)^{1/2}(a-kx'/2z'\xi)}$$
$$\times \exp\left(j\frac{\pi}{2}g^2\right) dg \left(\frac{\pi}{2\eta}\right)^{1/2} \int_{-(2\eta/\pi)^{1/2}(b+ky'/2z'\eta)}^{(2\eta/\pi)^{1/2}(b-ky'/2z'\eta)} \exp\left(j\frac{\pi}{2}h^2\right) dh \qquad (2\text{-}51)$$

These integrals have now been reduced to classical forms known as the

Fresnel integrals, which are defined by [8, 12]

$$C(x) = \int_0^x \cos\left(\frac{\pi t^2}{2}\right) dt \tag{2-52}$$

and

$$S(x) = \int_0^x \sin\left(\frac{\pi t^2}{2}\right) dt \tag{2-53}$$

and are tabulated in many handbooks [16]. Thus Eq. (2-51) can be written† as

$$\begin{aligned}
&\exp\left[-j\frac{k}{2z'}\left(\frac{x'^2}{(2z'/k)\xi}+\frac{y'^2}{(2z'/k)\eta}\right)\right]\frac{\pi}{2(\xi\eta)^{1/2}}\\
&\quad\times\left[C\left(\left(\frac{2\xi}{\pi}\right)^{1/2}\left(a-\frac{x'}{(2z'/k)\xi}\right)\right)+C\left(\left(\frac{2\xi}{\pi}\right)^{1/2}\left(a+\frac{x'}{(2z'/k)\xi}\right)\right)\right.\\
&\quad\left.+jS\left(\left(\frac{2\xi}{\pi}\right)^{1/2}\left(a-\frac{x'}{(2z'/k)\xi}\right)\right)+jS\left(\left(\frac{2\xi}{\pi}\right)^{1/2}\left(a+\frac{x'}{(2z'/k)\xi}\right)\right)\right]\\
&\quad\times\left[C\left(\left(\frac{2\eta}{\pi}\right)^{1/2}\left(b-\frac{y'}{(2z'/k)\eta}\right)\right)+C\left(\left(\frac{2\eta}{\pi}\right)^{1/2}\left(b+\frac{y'}{(2z'/k)\eta}\right)\right)\right.\\
&\quad\left.+jS\left(\left(\frac{2\eta}{\pi}\right)^{1/2}\left(b-\frac{y'}{(2z'/k)\eta}\right)\right)+jS\left(\left(\frac{2\eta}{\pi}\right)^{1/2}\left(b+\frac{y'}{(2z'/k)\eta}\right)\right)\right]
\end{aligned} \tag{2-54}$$

Before going on to see how this result affects Eq. (2-45), the arguments need to be examined in some detail. Namely, notice that $[(2z'/k)\xi]^{-1}$ and $[(2z'/k)\eta]^{-1}$ appear throughout the above terms. Using Eqs. (2-46) and (2-47), we notice that these terms become for large k‡

$$\left(\frac{2z'}{k}\xi\right)^{-1} = \left(1-\frac{2z'\beta}{k}\right)^{-1} \simeq 1+\frac{2z'\beta}{k} \tag{2-55a}$$

and

$$\left(\frac{2z'}{k}\eta\right)^{-1} = \left(1-\frac{2z'\gamma}{k}\right)^{-1} \simeq 1+\frac{2z'\gamma}{k} \tag{2-55b}$$

Further note that for the optical and quasi-optical regions, k is very large,

† Equation (2-54) was obtained by breaking Eq. (2-51) into the product of the sum of integrals each having limits of 0 to some finite value. To do this required another change of variables from g to $-g$ and h to $-h$ over the lower half range.

‡ Physically γ and β are like one over a radius of curvature which is of order z'.

making ξ and η very large. Thus, all of the arguments of the Fresnel functions are large, permitting the use of large-argument approximations for $C(\,)$ and $S(\,)$. For large arguments, the Fresnel functions are asymptotic† to 1/2. Using this approximation, the square bracketed terms in Eq. (2-54) reduce to

$$[C(\,) + C(\,) + jS(\,) + jS(\,)][C(\,) + C(\,) + jS(\,) + jS(\,)] \\ = (1 + j1)(1 + j1) = 2j \qquad (2\text{-}56)$$

Thus, Eq. (2-45) can be written as

$$u(x', y', z') = \frac{(2j)\pi}{2j\lambda z'\sqrt{\xi\eta}}\, e^{jkz'} \exp[-j(\beta x'^2 + \gamma y'^2)] \qquad (2\text{-}57)$$

The result can be further reduced by examining the coefficient out front of the exponential terms, and writing it as

$$\frac{1}{(2z'/k)(\xi\eta)^{1/2}} = \frac{1}{\{[(2z'/k)\xi][(2z'/k)\eta]\}^{1/2}} \\ = \frac{1}{[(1 - 2z'\beta/k)(1 - 2z'\gamma/k)]^{1/2}} \simeq 1 + \frac{z'}{k}(\beta+\gamma) \qquad (2\text{-}58)$$

Again, a term like $1 + 1/k$ appears, which in the limit of the short-wavelength approximation is approximately equal to one for large k when z' is of the order of $1/\beta$.

Hence Eq. (2-57) reduces to

$$u(x', y', z') = e^{jkz'} \exp[-j(\beta x'^2 + \gamma y'^2)] \qquad (2\text{-}59)$$

which is just the linear phase projection of the aperture quadratic phase function. Physically, this corresponds to the imaging of the aperture phase-transmittance function. This result applies to fields even quite close to the illuminating aperture since k is so large. Thus, without resorting to any a priori assumptions about the properties of lenses [11], an imaging characteristic has been derived for propagation in the Fresnel zone and beyond. Use will be made of these results in the chapter on lenses and imaging systems.

† We note that careful examination of the cornu spiral [16] shows that the asymptotic expansion of one-half is good for $x > 10$.

Problems

1. Verify that Eq. (2-29) yields an intensity function similar to Eq. (2-17).

2. Find the Fourier transform of the two functions

$$f(r) = \exp(-\beta r^2), \qquad g(r) = \exp(-j\beta r^2)$$

Hint: Consider using a technique of completing the square in the arguments.

3. Derive the Fresnel field for the rectangular aperture

$$u_t(x, y) = \begin{cases} A & -a < x < a, \quad -b < y < b \\ 0 & \text{otherwise} \end{cases}$$

Show that this rectangular aperture is just imaged as was the quadratic phase function in Eq. (2-59).

4. Find the Fraunhofer field of an aperture which has the phase function

$$u_t(x, y) = \begin{cases} e^{j(\pi/2)} & 0 < x < a \\ e^{-j(\pi/2)} & -a < x < 0 \end{cases} \quad \text{and zero otherwise}$$

5. Find the Fraunhofer field of three rectangular apertures separated a distance d and symmetric about the longitudinal axis

$$u_t(x, y) = \begin{cases} A & -d - b < y < -d + b, \quad -a < x < a \\ A & d - b < y < d + b, \quad -a < x < a \\ A & -b < y < b, \quad -a < x < a \\ 0 & \text{otherwise} \end{cases}$$

This result is similar to Young's double-slit experiment. Find the difference.

6. a. Find the Fraunhofer field for four rectangular apertures placed similar to Problem 5, except that

$$u_t(x, y) = \begin{cases} A & -2d - b < y < -2d + b, \quad -a < x < a \\ A & -d - b < y < -d + b, \quad -a < x < a \\ A & d - b < y < d + b, \quad -a < x < a \\ A & 2d - b < y < 2d + b, \quad -a < x < a \\ 0 & \text{otherwise} \end{cases}$$

Note that proper factoring of phase terms reduces the problem to Problem 5 with a new array factor.

b. Factor the result into two terms so that one term is the aperture factor shown in the text. Collect the phase terms into an appropriate array factor.

c. Factor Problem 5 in the same way as in Problem 6b.

d. What would you guess the array factor for six apertures to be like?

e. How about $2N$ apertures? Find a closed form.

Chapter III

PHYSICAL REALIZATIONS OF PHASE TRANSFORMERS, LENSES, AND SYSTEMS

Conceptual Lens

The description of how to realize some of the phase transformations used in Chapter II can now be given. To do so, an approximation will be employed that is commonly referred to as the "thin-lens approximation" [18, 19]. This approximation is equivalent to a statement that the emergent rays are nearly collinear with the incoming rays. That is, there is negligible transverse offset[†] δ of the ray as it enters on the first surface and exits on the last surface [8, 20]. Note that this does not say that the lens has to be physically thin. Some compound lenses obey this approximation.

In Chapter II it was found that input apertures, which have a quadratic phase-transmittance function, yield an output field that has the same functional form as the output field of the input aperture. This characteristic was called imaging and provided a useful property of image transfer. In this chapter the kind of surfaces that will provide such quadratic phase variations will be examined. Specifically, spherical and parabolic surfaces will be examined along with their associated phase-transmittance functions. From these transmittance functions the concept of focal length will be defined in terms of the surface curvatures.

To begin, note that essentially anything that is transparent and has an index of refraction different from the surrounding medium will produce a phase delay or retardation. Functionally, this retardation is represented by the exponent function $e^{j\phi(x,y)}$. The function $\phi(x, y)$ can be represented by

[†] Delta, δ, is shown in Fig. 3-1.

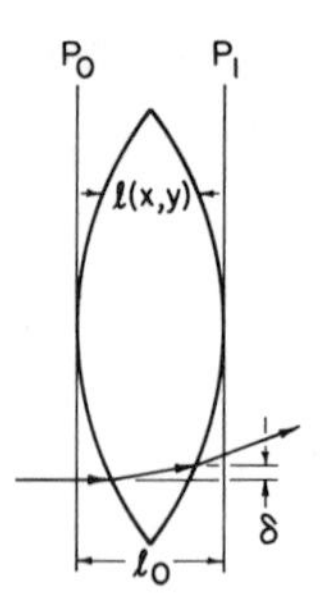

Fig. 3-1. General geometry for a lens showing the ray offset as δ.

a scalar product of the appropriate propagation vector **k** and path length $\mathbf{l}(x, y)$. Since a coordinate system that has the z axis aligned with the optical axis is assumed, this product can be written simply as

$$\phi(x, y) = kl_0 + k(\eta - 1)l(x, y) \tag{3-1}$$

where l_0 is the maximum thickness of the element, $k\eta l(x, y)$ is the phase delay or retardation in the lens element, and $k[l_0 - l(x, y)]$ is the remaining phase delay in the region bounded by the planes P_0 and P_1 shown in Fig. 3-1.

Spherical Surfaces

First, spherical surfaces will be examined. Utilizing rotational symmetry, these three-dimensional surfaces are described with two-dimensional pictures and the simpler equation of a circle.

First, however, a sign convention must be established. A positive radius of curvature will exist if the center of curvature is beyond the interface surface. A negative radius of curvature is associated with a system that has the center of curvature in a region in front of the interface surface. Figure 3-2 illustrates one case where R_1 is positive and R_2 is negative. Rays will be assumed to be moving from left to right.

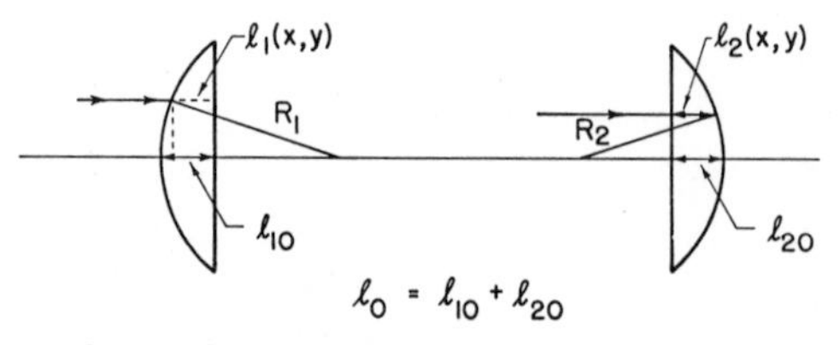

Fig. 3-2. Geometry for the lens equation derivation showing the lens divided along a plane of symmetry so that its surface curvatures appear separately.

In order to develop the equations describing $l(x, y)$, split the phase element into two parts. One surface of each part is flat, and the thickness at a particular transverse position (x, y) is denoted $l_1(x, y)$ and $l_2(x, y)$, respectively. Using the equation of a circle† and geometrical arguments, expressions can be written for $l_1(x, y)$ and $l_2(x, y)$ in terms of R_1, $-R_2$, x, and y as

$$l_1(x, y) = l_{10} - [R_1 - (R_1^2 - x^2 - y^2)^{1/2}] \tag{3-2a}$$

and

$$l_2(x, y) = l_{20} - [-R_2 - (R_2^2 - x^2 - y^2)^{1/2}] \tag{3-2b}$$

where the positive number $-R_2$ has been explicitly used to make our derivations independent of the sign of R_1 and R_2. If R_1 and $-R_2$ are factored out, Eqs. (3-2) become

$$l_1(x, y) = l_{10} - R_1[1 - [1 - (x^2 + y^2)/R_1^2]^{1/2}] \tag{3-3a}$$

and

$$l_2(x, y) = l_{20} + R_2[1 - [1 - (x^2 + y^2)/R_2^2]^{1/2}] \tag{3-3b}$$

Using the para-axial approximation

$$x^2 + y^2 \ll R_1^2 \quad \text{or} \quad R_2^2 \tag{3-4}$$

and the binomial expansion on the square-root terms leaves

$$l_1(x, y) = l_{10} - (x^2 + y^2)/2R_1 \tag{3-5a}$$

and

$$l_2(x, y) = l_{20} + (x^2 + y^2)/2R_2 \tag{3-5b}$$

When these two equations are combined together, an element thickness at a position (x, y) is given by

$$l(x, y) = l_0 - \tfrac{1}{2}(1/R_1 - 1/R_2)(x^2 + y^2) \tag{3-6}$$

If Eq. (3-1) and Eq. (3-6) are combined, an expression for the phase-transmittance function $T(x, y)$ can be written as

$$T(x, y) = \exp[jk\eta l_0 - (k/2)(\eta - 1)(1/R_1 - 1/R_2)(x^2 + y^2)] \tag{3-7}$$

Following the classical approach [8, 11, 18, 20], a focal length f can be

† In this case it is $z^2 + \varrho^2 = R_i^2$, where $\varrho^2 = x^2 + y^2$.

defined for a simple lens as

$$1/f = (\eta - 1)(1/R_1 - 1/R_2) \tag{3-8}$$

If one applies this definition to other surface curvature arrangements, the signs of R_1 and R_2 must be established in accordance with the convention outlined earlier. In this case then, the formula for the focal length of a double-convex lens would be

$$1/f = (\eta - 1)(1/R_1 + 1/R_2) \tag{3-9}$$

where R_2 is recognized to be negative because its center of curvature lies ahead of the surface in the sense that the rays are traveling from left to right.

Thus the phase-transmittance function of a lens can be written as

$$T(x, y) = \exp[jk\eta l_0 - j(k/2f)(x^2 + y^2)] \tag{3-10}$$

This result can be compared with the results of Chapter II, Eqs. (2-43) and (2-46), if β and γ are replaced by $k/2f$, where the constant phase term $(jk\eta l_0)$ is neglected.

Application of the imaging concept from Chapter II indicates that spherical surfaces of appropriate radii of curvature can form images. It is important to note that negative focal-length lenses can be formed by careful selection of the radii of curvature in both sign and in relative magnitude.

Figures 3-3 and 3-4 illustrate several cases of lens configurations along with the corresponding focal-length sign conventions. The reader should verify for each case that the focal-length signs are as indicated.

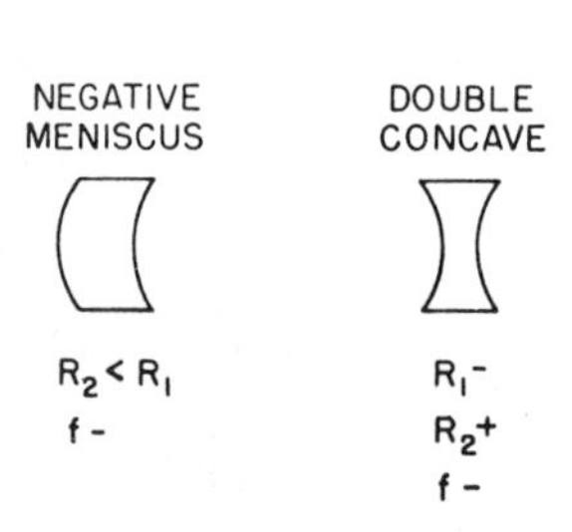

Fig. 3-3. Sketches of negative focal-length lenses.

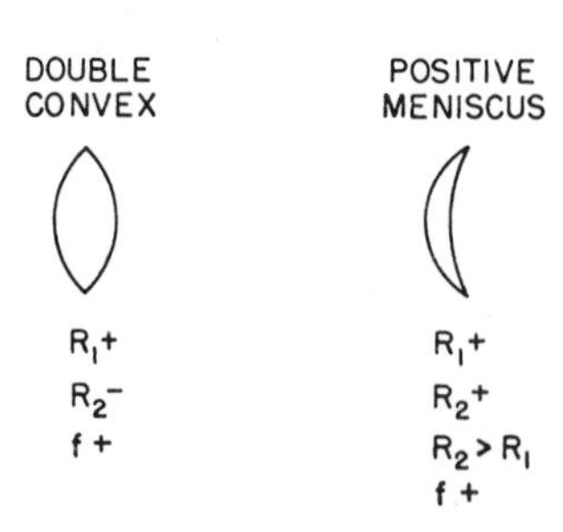

Fig. 3-4. Sketches of positive focal-length lenses.

Parabolic Surfaces

Lenses can be generated by using parabolic surfaces, if the following equation is used to describe the surface illustrated in Fig. 3-5.

$$z = A - B(x^2 + y^2) \tag{3-11}$$

Following the methods for spherical surfaces, we can write the element thickness $l(x, y)$ as

$$l(x, y) = (A - A') - 2B(x^2 + y^2) \tag{3-12}$$

This expression can be used to define a phase-transmittance function like Eq. (3-10) if a focal length is defined as

$$1/f = 4(\eta - 1)B \tag{3-13}$$

This then enables a clear correspondence with the description of how to use spherical surfaces as lens elements.

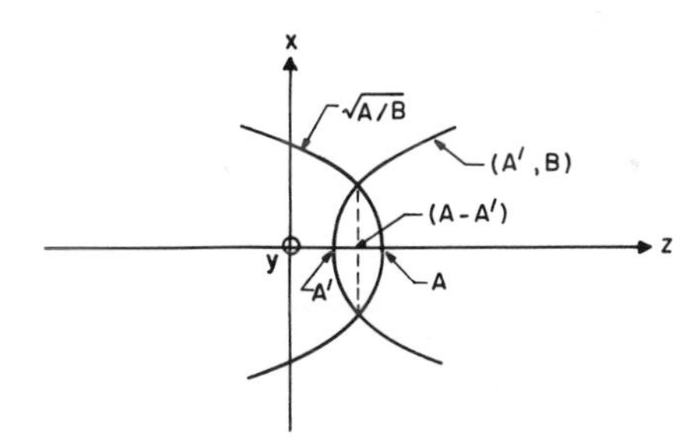

Fig. 3-5. Geometry for a parabolic surface approximation for a lens.

In retrospect, the approximation used in the spherical-surface analysis makes the use of parabolic surfaces an obvious candidate for a lens element. Thus the parabolic surface is a close approximation to the desired surface that would provide the needed quadratic-phase variation.

In addition, the use of large spherical surfaces can lead to errors in the approximation of the quadratic-phase function. These errors arise, because the para-axial assumption is not very good. These errors appear in some lens elements and are called aberrations. That is, focal points of the outer periphery of the lens will not be coincident with the foci formed by the central portion of the lens.

Systems of Lenses

In Chapter II the form of far fields or receiving aperture fields was established for the case when the input apertures have various kinds of phase-transmittance functions. This chapter began with a derivation of the physical form these phase devices would take in order to have the needed quadratic variation in transverse phase. To this point then, the analysis has been restricted to single-system elements and their effects. It is now necessary to examine how combinations of lenses and lens systems would work together to perform specific tasks, such as imaging, Fourier transforming, or some other optical-processing functions.

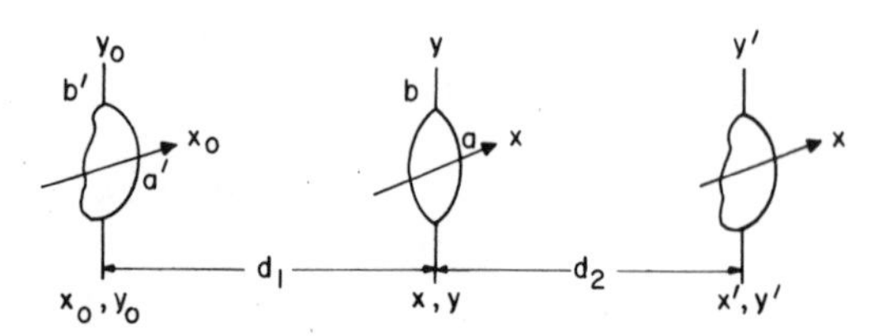

Fig. 3-6. A simple optical system having one lens and input/output planes.

To begin, a simple lens system like the one shown in Fig. 3-6 is examined. The problem is formulated by representing the output fields as functions of the input fields in each stage. This can be viewed as an organized backward approach. Thus the field in aperture 3 due to aperture 2 is written as

$$u(x', y', d_2) = \frac{\exp(jkd_2)}{j\lambda d_2} \exp\left[j \frac{k}{2d_2} (x'^2 + y'^2)\right]$$
$$\times \iint_{-\infty}^{\infty} u_t(x, y) P(x, y) \exp\left[j \frac{k}{2d_2} (x^2 + y^2)\right]$$
$$\times \exp\left[-j \frac{k}{d_2} (xx' + yy')\right] dx\, dy \tag{3-14}$$

where $P(x, y)$ is called the pupil function for the lens aperture. It is the function that gives a finite size to the lens. In general, $P(x, y)$ is represented as

$$P(x, y) = \begin{cases} 1 & -a < x < a, \quad -b < y < b \\ 0 & \text{otherwise} \end{cases} \tag{3-15}$$

Now $u_t(x, y)$ is the field transmitted by the lens aperture. To write this in terms of incident field and the aperture transmittance function, the following form is used:

$$u_t(x, y) = u_i(x, y)T(x, y) \tag{3-16}$$

where $T(x, y)$ is the transmittance function of a lens,

$$T(x, y) = \exp\left[-j\frac{k}{2f}(x^2 + y^2)\right] \tag{3-17}$$

Note that the product $T(x, y)P(x, y)$ gives a truncated lens equation which defines a lens pupil. Now $u_i(x, y)$ is defined by an equation similar to Eq. (3-14). That is, for the field in aperture 2 a field is excited by aperture 1,

$$\begin{aligned} u_i(x, y, d_1) = {} & \frac{\exp(jkd_1)}{j\lambda d_1} \exp\left[j\frac{k}{2d_1}(x^2 + y^2)\right] \\ & \times \iint_{-\infty}^{\infty} u(x_0, y_0) \exp\left[j\frac{k}{2d_1}(x_0^2 + y_0^2)\right] \\ & \times \exp\left[-j\frac{k}{d_1}(xx_0 + yy_0)\right] dx_0\, dy_0 \end{aligned} \tag{3-18}$$

Combining these equations, a double set of double integrals is obtained in the form

$$\begin{aligned} u(x', y', d_0) = {} & -\frac{\exp[jk(d_1 + d_2)]}{\lambda^2 d_1 d_2} \exp\left[j\frac{k}{2d_2}(x'^2 + y'^2)\right] \\ & \times \iint_{-\infty}^{\infty} \exp\left[j\frac{k}{2}(x^2 + y^2)\left(\frac{1}{d_1} + \frac{1}{d_2}\right)\right] T(x, y)P(x, y) \\ & \times \left[\iint_{-\infty}^{\infty} u(x_0, y_0) \exp\left[j\frac{k}{2d_1}(x_0^2 + y_0^2)\right]\right. \\ & \left. \times \exp\left[-j\frac{k}{d_1}(xx_0 + yy_0)\right] dx_0\, dy_0\right] \\ & \times \exp\left[-j\frac{k}{d_2}(xx' + yy')\right] dx\, dy \end{aligned} \tag{3-19}$$

where d_0 equals $d_1 + d_2$. This is a general result relating the sequential fields in the system by a general integral equation. To obtain some understanding of this equation, rearrange and substitute Eqs. (3-15) and (3-17) into Eq. (3-19), leaving

$$u(x', y', d_0) = \left[-\frac{\exp[jk(d_1 + d_2)]}{\lambda^2 d_1 d_2} \exp\left[j\frac{k}{2d_2}(x'^2 + y'^2) \right] \right]$$
$$\times \iint_{-\infty}^{\infty} dx_0\, dy_0 \int_{-b}^{b} \int_{-a}^{a} dx\, dy$$
$$\times \left[\exp\left[j\frac{k}{2}\left(\frac{1}{d_1} + \frac{1}{d_2} - \frac{1}{f} \right)(x^2 + y^2) \right] \right.$$
$$\times \left. \exp\left\{ -jk\left[\left(\frac{x_0}{d_1} + \frac{x'}{d_2} \right)x + \left(\frac{y_0}{d_1} + \frac{y'}{d_2} \right)y \right] \right\} \right]$$
$$\times \left[u_0(x_0, y_0) \exp\left[j\frac{k}{2d_1}(x_0^2 + y_0^2) \right] \right] \tag{3-20}$$

A direct general integration over the lens aperture variables (x, y) is not easy. This difficult problem is approached by considering some special cases. First, examine the imaging case by arranging output aperture spacing such that the coefficient of the quadratic term over (x, y) is forced to zero,

$$1/d_1 + 1/d_2 - 1/f = 0 \tag{3-21}$$

This result should not be surprising since it is the simple *lens law* [19] that is familiar from imaging with simple lenses. It is more commonly written as

$$1/d_o + 1/d_i = 1/f \tag{3-22}$$

To reduce the number of parameters, use the constraint defined by Eq. (3-22) and write either d_1 or d_2 as a function of d_2 or d_1 and f. For purposes of this illustration, write

$$d_1 = d_2 f/(d_2 - f) \tag{3-23}$$

In this case, the quadratic term under the (x, y) integral goes to unity leaving a straightforward integration over x and y. Thus,

$$u(x', y', d_0) = [\;] \iint_{-\infty}^{\infty} dx_0\, dy_0\, u(x_0, y_0) \exp\left[j\frac{k}{2}\left(\frac{d_2 - f}{d_2 f} \right)(x_0^2 + y_0^2) \right]$$
$$\times \left\{ 4ab \operatorname{sinc}\left(ka\left[\left(\frac{d_2 - f}{d_2 f} \right)x_0 + \frac{x'}{d_2} \right] \right) \right.$$
$$\times \left. \operatorname{sinc}\left(kb\left[\left(\frac{d_2 - f}{d_2 f} \right)y_0 + \frac{y'}{d_2} \right] \right) \right\} \tag{3-24}$$

is the general result.[†] Some very interesting cases are illustrated by this result. First, examine the size of ka and kb. Both of these quantities are of the order of a/λ or b/λ, and since

$$b, a \gg \lambda \tag{3-25}$$

the sinc functions can be approximated by delta functions,[‡] leaving

$$u(x', y', d_0) = [\;]\, 4 \iint_{-\infty}^{\infty} dx_0\, dy_0\, u(x_0, y_0) \exp\left[j \frac{k}{2}\left(\frac{d_2 - f}{d_2 f}\right)(x_0^2 + y_0^2)\right]$$
$$\times \left(\frac{\lambda}{2}\right)^2 \delta\left[x_0\left(\frac{d_2 - f}{d_2 f}\right) + \frac{x'}{d_2}\right] \delta\left[y_0\left(\frac{d_2 - f}{d_2 f}\right) + \frac{y'}{d_2}\right] \tag{3-26}$$

Now Eq. (3-26) is an integral over (x_0, y_0) involving delta functions. From the work in Chapter I this greatly simplifies the integral, giving[§]

$$u(x', y', d_0) = -\left[\exp\left(j \frac{k d_2^2}{d_2 - f}\right)\right]\left(\frac{f}{d_2 - f}\right) \exp\left[j \frac{k(x'^2 + y'^2)}{2(d_2 - f)}\right]$$
$$\times u_0\left[-\left(\frac{f}{d_2 - f}\right)x', -\left(\frac{f}{d_2 - f}\right)y'\right] \tag{3-27}$$

where u_0 has been used to denote the exciting aperture field. The minus signs in the argument of $u_0(\;)$ arise from the delta-function operator.

Now examine some of the implications of this result. The fact that $u(x', y', d_0)$ is directly proportional to $u_0(\;)$ implies that direct imaging has occurred. Further, the minus signs indicate that there has been an inversion of the image, and the factor $f/(d_2 - f)$ indicates that there has been a magnification. Therefore $(d_2 - f)/f$ is referred to as the magnification of the system α,

$$\alpha \equiv (d_2 - f)/f = x'/x_0 \tag{3-28}$$

The field in the observation aperture is reproduced except for the quadratic-phase error and the attenuation factor $(f/d_2 - f)$.

If the special case of $d_2 = 2f$ or unit magnification is considered, the phase error disappears, and a simple inverted image with the characteristic phase

[†] Here a certain notation for the sinc function, defined by [21],

$$\sin(ax)/ax = \operatorname{sinc}(ax)$$

is being used.

[‡] Here we are defining [21] the delta functions by $\delta(x) = \lim_{g \to \infty} [\sin(gx)/\pi x]$.

[§] To do this we need the following identity, $\delta(ax) = (1/a)\,\delta(x)$.

error is obtained. This result is simply an inverted image

$$u(x', y', 4f) = -(e^{jk4f})\left\{\exp\left[-j\frac{k}{4f}(x'^2 + y'^2)\right]\right\}u_0(-x', -y') \tag{3-29}$$

It illustrates that the functional form of the field is reproduced with a multiplicative phase term corresponding to the separation distance of the two aperture field points.

Another result inherent in Eq. (3-27) is that the amplitude coefficient outside is proportional to $1/d_2$, making the output wave have the form of a spherical wave. For $d_2 < f$, it is a converging spherical wave, and for $d_2 > f$ it is a diverging wave. The reader can easily confirm this by drawing a plane wave or collimated beam incident onto a simple lens and showing the appropriate focus at $d_2 = f$. This is an important result showing that lens systems can be used to transmit information or image data through a system [22–24].

Fourier Transform

In the case that the imaging constraint defined by Eq. (3-22) is not utilized, the integral in Eq. (3-20) can be rearranged to form a product of two integrals. To illustrate, consider Eq. (3-20) after rearranging the terms in the second bracket to

$$\begin{aligned} u(x', y', d_0) = [\ ] \iint dx_0\, dy_0\, u(x_0, y_0) \exp\left[j\frac{k}{2d_1}(x_0{}^2 + y_0{}^2)\right] \\ \times \left\{\int_{-a}^{a} dx \exp\{j\zeta[x^2 - \xi(x_0, x')x]\}\right. \\ \left. \times \int_{-b}^{b} dy \exp\{j\zeta[y^2 - \xi(y_0, y')y]\}\right\} \end{aligned} \tag{3-30}$$

where

$$\zeta = \frac{k}{2}\left(\frac{1}{d_1} + \frac{1}{d_2} - \frac{1}{f}\right) \tag{3-31a}$$

$$\xi(x_0, x') = \frac{k}{\zeta}\left(\frac{x_0}{d_1} + \frac{x'}{d_2}\right) \tag{3-31b}$$

and

$$\xi(y_0, y') = \frac{k}{\zeta}\left(\frac{y_0}{d_1} + \frac{y'}{d_2}\right) \tag{3-31c}$$

If the square is completed in the exponent of the bracketed terms, the term in the large braces can be written as

$$\{\,\} = \exp\left\{-j\frac{\zeta}{4}\,[\xi^2(x_0, x') + \xi^2(y_0, y')]\right\}$$
$$\times \int_{-a}^{a} dx\,\exp\{j\zeta[x - \xi(x_0, x')]^2\} \int_{-b}^{b} dy\,\exp\{j\zeta[y - \xi(y_0, y')]^2\} \tag{3-32}$$

With appropriate changes in variables and rearrangement of terms, Eq. (3-32) becomes just like Eq. (2-52) and, under the assumption that ka and kb are very much larger than unity, becomes

$$\{\,\} = \frac{\pi j}{\zeta}\exp\left\{-j\frac{\zeta}{4}\,[\xi^2(x_0, x') + \xi^2(y_0, y')]\right\} \tag{3-33}$$

If Eqs. (3-31b) and (3-31c) are substituted, the term in the braces is

$$\{\,\} = \frac{\pi j}{\zeta}\exp\left\{-j\frac{k^2}{4\zeta}\left[\frac{x_0{}^2 + y_0{}^2}{d_1{}^2} + \frac{x'^2 + y'^2}{d_2{}^2} + \frac{2}{d_1 d_2}(x_0 x' + y_0 y')\right]\right\} \tag{3-34}$$

which reduces Eq. (3-30) to

$$u(x', y', d_0) = \left[\frac{jf}{\lambda}\,\frac{\exp[jk(d_1 + d_2)]}{d_2 f + d_1 f - d_1 d_2}\right.$$
$$\left.\times \exp\left[j\frac{k}{2}(x'^2 + y'^2)\left(\frac{f - d_1}{d_2 f + d_1 f - d_1 d_2}\right)\right]\right]$$
$$\times \iint_{-\infty}^{a} dx_0\,dy_0\,u(x_0, y_0)$$
$$\times \exp\left[j\frac{k}{2}(x_0{}^2 + y_0{}^2)\left(\frac{f - d_2}{d_2 f + d_1 f - d_1 d_2}\right)\right]$$
$$\times \exp\left[-j\frac{kf}{d_2 f + d_1 f - d_1 d_2}(x_0 x' + y_0 y')\right] \tag{3-35}$$

Notice that this equation seems very complicated with quadratic-phase terms inside and outside the integrals. Considerable simplification results, however, when the input and output planes are spaced a focal distance away from the lens. Thus when d_1 and d_2 are both equal to f, Eqs. (3-35) and (3-30) simplify to

$$u(x', y', d_0) = \frac{j}{\lambda f}\exp(jkd_0)\iint_{-\infty}^{\infty} dx_0\,dy_0\,u(x_0, y_0)$$
$$\times \exp\left[-j\frac{k}{f}(x_0 x' + y_0 y')\right] \tag{3-36}$$

which is just the form of the Fourier transform [13, 23, 25]. Generally the constant phase terms in front of the integral are neglected leaving a very simple form expressing

$$u(x', y', d_0) = \mathscr{F}\{u(x_0, y_0)\} \tag{3-37}$$

where $\mathscr{F}\{\ \}$ represents the two-dimensional Fourier transform.

The two cases just detailed illustrate the two principal mathematical results obtainable with lenses. From the work in Chapter II, it was clear that it would be possible to obtain these results. Now, however, it is possible to write a "two-terminal" system representation of optical systems where aperture S_0 represents the input and aperture S' represents the output. In this system context then, image transformation, such as Fourier transforming, or simple image transfer can be done with possible scale and amplitude changes. In general, a linear system representation for a simple optical system has been derived.

Problems

1. Verify that the positive focal-length lenses as shown in Fig. 3-4 are indeed positive for the curvature conditions indicated.

2. There are two lenses not shown in Fig. 3-4 which are similar to those presented. What are they? *Hint*: They arise from an extrema condition on R_2.

3. Verify by simple ray drawings that the sign convention for f indeed places a focal point either in the region already traversed (−) or beyond the surfaces (+).

4. Examine the image inversion for the case that $d_2 < f$ and $d_2 > f$. Make a ray-optic sketch.

5. Perform a similar analysis, as done for Eq. (3-17) on imaging, for the case that

$$T(x, y) = \exp\left(-j\frac{k}{2f}x^2\right)$$

6. Repeat Problem 5 for the case that

$$T(x, y) = \exp[-j(\beta x + \gamma y)]$$

7. Starting with Eqs. (3-30), (3-31), and (3-32), use methods similar to that in Chapter II [Eq. (2-52)] to derive Eq. (3-35).

Chapter IV

SYSTEM TRANSFORM CONCEPTS AND NOTATION

Fourier Transform Representation of Fields

Thus far it has been established that electromagnetic radiation from one aperture to the next can be represented by an integral equation called the Rayleigh–Sommerfeld equation. It is a Fredholm integral equation like the Fourier transform. In Chapter II the details of how one manipulates the integral equation, using various approximations, into an explicit form of the Fourier transform are given. In Chapter III the physical definition of the lens elements was derived along with the mathematical details of how to arrange an optical system that would perform the operations of imaging and Fourier transforming.

The fact that a linear system description is available for relating the input and output fields of an optical system suggests that both the input and output fields can be described in terms of appropriate Fourier spectra. The use of Fourier spectra along with the simple kernel e^{jux} as the expansion basis provides a framework for physically describing this spectrum representation.

Fourier spectrum representation physically is equivalent to expressing the fields in a region as a superposition of plane waves. Waves emanating from apertures do so at many angles. These angles are each representable as components in a spectral expansion. Thus there is a unique angular spatial spectrum associated with each aperture and its exciting field. A single plane wave, for example, is a single component in a spectral representation.

Further, it is known from circuit analysis that narrower and narrower time pulses require more and more bandwidth to transmit the complete description. Similarly, smaller and smaller apertures cause the output wave to have more and more angular spectrum.

To elucidate this discussion, let us assume that the wave emanating from an aperture can be decomposed into a sum of plane waves by expanding it in a Fourier transform [13, 23] as follows:

$$V(u, v, z) = \iint_{-\infty}^{\infty} U(x, y, z) \exp[-j(ux + vy)]\, dx\, dy \tag{4-1}$$

where u and v are spatial frequency variables that have dimension of inverse length or cycles per meter. In general, the association of u and x, and v and y will be consistent throughout.

The existence of the integral, Eq. (4-1), is assured by the fact that the fields satisfy the radiation condition, i.e., arise from sources of finite energy. Mathematically, the functions must be piecewise smooth (finite number of discontinuities) and the following integrals must exist [26]:

$$\int_{-\infty}^{\infty} | U(x, y, z) |\, dx < \infty \tag{4-2a}$$

and

$$\int_{-\infty}^{\infty} | U(x, y, z) |\, dy < \infty \tag{4-2b}$$

If Eq. (4-1) exists, then a corresponding inverse transform can be written as

$$U(x, y, z) = \left(\frac{1}{2\pi}\right)^2 \int_{-\infty}^{\infty} \int_{-\infty}^{\infty} V(u, v, z) \exp[j(ux + vy)]\, du\, dv \tag{4-3}$$

Notice that Eq. (4-3) is nearly symmetric in the (u, v) variables except that there is the normalization constant $(1/2\pi)^2$, and the change in sign of the exponent.

Equation (4-1) and (4-3) form what is known as a transform pair and can be used to form alternate representations of the field in a region.† The angular spectrum of the wave $U(x, y, z)$ is referred to as $V(u, v, z)$. It is a complex function, so it has an amplitude and a phase. The form of this alternate representation for the fields in the region can be found by examining what happens to the wave equation under the transformation.

† The idea of alternate representations of a field should not be foreign since this is done in waveguides by using TM and TE modes or alternately of having plane waves at different angles.

Transform of the Wave Equation

The wave equation transforms by substituting Eq. (4-3) into

$$(\nabla^2 + k^2)U(x, y, z) = 0 \tag{4-4}$$

yielding

$$\left(\frac{\partial^2}{\partial x^2} + \frac{\partial^2}{\partial y^2} + \frac{\partial^2}{\partial z^2} + k^2\right)\left(\frac{1}{2\pi}\right)^2 \int_{-\infty}^{\infty} \int_{-\infty}^{\infty} V(u, v, z) \times \exp[j(ux + vy)]\, du\, dv = 0 \tag{4-5}$$

Now the only term under the integral that depends on x and y is the exponent term. Thus the differential operators can be commuted with the integral operators, allowing the x and y differentiation to be performed under the integral sign, giving

$$\left(\frac{1}{2\pi}\right)^2 \int_{-\infty}^{\infty} \int_{-\infty}^{\infty} \left[\frac{\partial^2 V}{\partial z^2} + (k^2 - u^2 - v^2)V\right] \exp[j(ux + vy)]\, du\, dv = 0 \tag{4-6}$$

In order for this integral to be zero for all points (x, y, z), the bracketed term in the integrand must be zero, leaving

$$\left[\frac{\partial^2}{\partial z^2} + (k^2 - u^2 - v^2)\right]V = 0 \tag{4-7}$$

Since this is a simple second-order differential equation over z, the solution is known to be of the following form:

$$V(u, v, z) = A^+(u, v) \exp[j(k^2 - u^2 - v^2)^{1/2} z] + A^-(u, v) \exp[-j(k^2 - u^2 - v^2)^{1/2} z] \tag{4-8}$$

This representation is just that of forward and reflected plane waves, where the A^+ and A^- represent the coefficients for the forward and reverse waves, respectively. As before, only the forward waves will be considered as emanating from the aperture, therefore†

$$V(u, v, z) = A(u, v) \exp[j(k^2 - u^2 - v^2)^{1/2} z] \tag{4-9}$$

The investigation of this alternate representation of fields in the region began by examining the transformation of the wave equation. The trans-

† It is important to keep in mind that propagation only occurs for $u^2 + v^2 < k^2$; otherwise, the exponent term will represent an attenuation.

formed field expression is then associated with how the spectrum propagates in the region. However, to make a meaningful comparison the transform of the aperture field $U(x, y, 0)$ should be known. This can be written as

$$V(u, v, 0) = \iint_{-\infty}^{\infty} U(x, y, 0) \exp[-j(ux + vy)]\, dx\, dy \qquad (4\text{-}10)$$

Note, however, that either representation must be at least piecewise continuous and since $U(x, y, z)$ is continuous, so must $V(u, v, z)$ be. In that case then

$$\begin{aligned} V(u, v, 0) &= V(u, v, z)\,|_{z=0} \\ &= A(u, v) \exp[j(k^2 - u^2 - v^2)^{1/2} z]\,|_{z=0} \end{aligned} \qquad (4\text{-}11)$$

leaving

$$V(u, v, 0) = A(u, v) \qquad (4\text{-}12)$$

Therefore, the coefficient $A(u, v)$ is just the spectrum of the aperture field, and it propagates in the region $z > 0$ with additive phase. This is represented by multiplication with the phase factor $\exp[j(k^2 - u^2 - v^2)^{1/2} z]$. This result is quite interesting since it says that spectrums propagate and are represented at any point (x, y, z) by keeping track of the linear phase terms.

This representation is quite compatible with the ray-optic picture which says that rays correspond to phase fronts moving through a region and that only changes in phase need to be accounted for by use of the Eikonal equation or Fermat's principle. Thus the plane-wave representation in the form of the Fourier transform is very analogous to the geometrical optics picture.

Secondly, this method affords a nice way of accounting for what happens to the spectrum of an illuminating aperture once it propagates into a region. This approach can be used to examine what happens to a plane wave that encounters an aperture.

Propagation as a Transfer Function

Having formulated the Fourier transform at any point z in the region of propagation, the transform can be related at z to the transform at $z = 0$. Using Eqs. (4-9) and (4-12), we can form the transfer ratio of the transforms

$$\frac{V(u, v, z)}{V(u, v, 0)} = \exp[i(k^2 - u^2 - v^2)^{1/2}z] \tag{4-13}$$

which is really the transfer function of the region for propagation.

If $T(u, v, z)$ is the transfer function, then a region of applicability can be defined as

$$T(u, v, z) = \frac{V(u, v, z)}{V(u, v, 0)} = \begin{cases} \exp[i(k^2 - u^2 - v^2)^{1/2}z] & u^2 + v^2 < k^2 \\ 0 & u^2 + v^2 \geq k^2 \end{cases} \tag{4-14}$$

where $u^2 + v^2 = k^2$ defines a circle in the transverse (u, v) plane. Inside this circle there is propagation with modulus equal to one. Outside this circle there is attenuation or evanescent waves. Although this result is exceedingly simple, it is a very useful concept, and one which will be used later in the work with optical devices used as linear circuit elements. Since it has been shown that propagation in a region can be viewed as an operation with a linear filter [25–29], other devices can be investigated and analyzed as filters. This will be covered in more detail later, but it is useful at this point to see the mathematical preliminaries.

If an incident wave $U(x, y, z')$ illuminated a structure located at z_0, then the field emanating from the structure can be written as

$$U(x, y, z_0)t(x, y, z_0) = U_t(x, y, z_0) \tag{4-15}$$

Examples of $t(x, y, z_0)$ might be:

(a) A rectangular aperture

$$t(x, y, z_0) = \begin{cases} 1 & -a < x < a, \quad -b < y < b, \quad z = z_0 \\ 0 & x > |a|, \quad y > |b| \end{cases} \tag{4-16}$$

(b) Two rectangular apertures on a diagonal

$$t(x, y, z_0) = \begin{cases} 1 & -2a < x < -a, \quad a < x < 2a, \\ & -2b < y < -b, \quad b < y < 2b \\ 0 & x \text{ and } y \text{ elsewhere} \end{cases} \tag{4-17}$$

(c) Or n apertures as shown in Fig. 4-1.

Since the spatial field emanating from a structure is written as a product of the structure function and the incident field, convolution [25, 27] is used if the transforms of these functions are being considered, i.e.,

$$V_t(u, v, z_0) = V(u, v, z_0) * \tau(u, v, z_0) \tag{4-18}$$

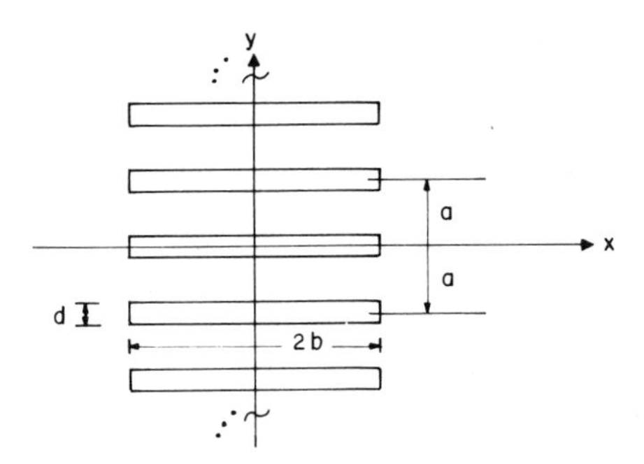

Fig. 4-1. Geometry of a sequence of N apertures centered about the origin along the y axis.

where $\tau(u, v, z_0)$ is the transverse Fourier transform of $t(x, y, z_0)$. Formally, this can be written as

$$V * \tau = \int_{-\infty}^{\infty} \int_{-\infty}^{\infty} V(\xi, \eta, z_0)\tau(u - \xi, v - \eta, z_0)\, d\xi\, d\eta \qquad (4\text{-}19)$$

The mathematical basis of this concept has been established. Further details will follow after more theoretical development.

General Transform Relationships of Propagation

In Chapter II it was shown how the field anywhere in a region is related by an integral equation to the field of an exciting aperture. That equation in general can be written as

$$U(x', y', z') = \iint_{-\infty}^{\infty} U_0(x, y, z)h(x, y, z; x', y', z')\, dx\, dy \qquad (4\text{-}20)$$

where $h(\ )$ took on a special form

$$h(x, y, 0, z', y', z') = \frac{e^{jkz'}}{j\lambda z'} \exp\left\{j \frac{k}{2z'} [(x' - x)^2 + (y' - y)^2]\right\} \qquad (4\text{-}21)$$

If Eq. (4-21) is substituted into Eq. (4-20), we have

$$U(x', y', z') = \frac{e^{jkz'}}{j\lambda z'} \iint_{-\infty}^{\infty} U_0(x, y)$$
$$\times \exp\left\{j \frac{k}{2z'} [(x' - x)^2 + (y' - y)^2]\right\} dx\, dy \qquad (4\text{-}22)$$

Referring to Eq. (4-15), we see that this equation is now in the form of a convolution multiplied by a complex constant p, where

$$p(z') = \frac{e^{jkz'}}{j\lambda z'} \tag{4-23}$$

Using this notation, write Eq. (4-22) as

$$U_1(x', y') = U_0 * h \tag{4-24}$$

where $U(x', y', z') = p(z')U_1(x', y')$.

Alternatively write the Fourier transform equivalent of this convolution as

$$F\{U_1\} = F\{U_0\}F\{h\} \tag{4-25}$$

or†

$$V(u, v, z') = V_0(u, v)H(u, v)p(z') \tag{4-26}$$

The form of $V_0(u, v)$ was shown earlier to be the convolution of the field spectrum illuminating the excitation aperture and the spectrum of the aperture, which in the case of a plane wave was a delta function convolved with the aperture spectrum $T(u, v, z)$.‡ Thus write

$$V_0(u, v) = V_i(u, v) * T(u, v) \tag{4-27}$$

which for a normally incident plane wave is just

$$V_0(u, v) = T(u, v) \tag{4-28}$$

It remains to examine $H(u, v)$. The function $h(x', y', z'; x, y)$ is written as the shifted term in the convolution. Since it is a space-invariant function, $H(u, v)$ can be found by finding the explicit transform of

$$h(x, y, z') = \frac{e^{jkz'}}{j\lambda z'} \exp\left[j \frac{k}{2z'} (x^2 + y^2)\right] \tag{4-29}$$

Using a general form of Eq. (4-1) gives

$$H(u, v, z') = \frac{e^{jkz'}}{j\lambda z'} \iint_{-\infty}^{\infty} \exp\left[j \frac{k}{2z'} (x^2+y^2) - j(ux+vy)\right] dx\, dy \tag{4-30}$$

† $V = F\{U\}$, $V_0 = F\{U_0\}$, and $H = F\{h\}$.

‡ See Eq. (4-14).

This integration is done by first completing the square and rearranging, giving

$$H(u, v, z') = \frac{e^{jkz'}}{j\lambda z'} \exp\left[-j \frac{z'}{2k} (u^2 + v^2)\right] \times \iint_{-\infty}^{\infty} \exp\left\{j \frac{k}{2z'} \left[\left(x - \frac{z'u}{k}\right)^2 + \left(y - \frac{z'y}{k}\right)^2\right]\right\} dx\, dy \tag{4-31}$$

Using techniques similar to those used in Chapter I, the double integral can be done by a simple change of variable, obtaining [16]

$$\iint_{-\infty}^{\infty} \exp\left\{j \frac{k}{2z'} \left[\left(x - \frac{z'u}{k}\right)^2 + \left(y - \frac{z'v}{k}\right)^2\right]\right\} dx\, dy = j\lambda z' \tag{4-32}$$

which gives for $H(u, v, z')$

$$H(u, v, z') = e^{jkz'} \exp\left[-j \frac{z'}{2k} (u^2 + v^2)\right] \tag{4-33}$$

This result can be verified by examining the general transfer function, Eq. (4-4), for the region

$$u, v < k \tag{4-34}$$

In that case

$$T(u, v, z') = \exp[j(k^2 - u^2 - v^2)^{1/2} z'] \qquad u^2 + v^2 < k^2 \tag{4-35}$$

which can be approximated by

$$T(u, v, z') = e^{jkz'} \exp\left[-j \frac{z'}{2k} (u^2 + v^2)\right] \qquad u^2 + v^2 \ll k^2 \tag{4-36}$$

where the binomial expansion was used on the radical. Equation (4-36) checks very nicely with Eq. (4-33) and verifies our approach. This result applies explicitly to the Fresnel zone.

If one were to examine the appropriate $h(\ )$ function for the Fraunhofer zone, a problem would arise in that the spatial invariance was explicity destroyed by the Fraunhofer approximation. This seems to imply that no transfer function exists for the Fraunhofer zone; however, the more general result of the Fresnel zone applies to the more restrictive Fraunhofer region, thus validating the use of $H(u, v)$ in the Fraunhofer zone by analytic continuation.

The important point is that now the propagation within a system can be represented by multiplication with a certain transform function $H(u, v)$. For the sake of the later work in operational notation, the z' dependence will be dropped. Thus one must always remember that the total phase delay of the system has to have the linear z'-dependent phase term added in.

This short-hand expression for propagation will be designated as

$$Q(u, v) = \exp\left[-j \frac{z'}{2k} (u^2 + v^2)\right] \tag{4-37}$$

Now it is quite easy to show that the corresponding spatial function that corresponds to $Q(u, v)$ is

$$Q_1(x, y) = \frac{1}{j\lambda z'} \exp\left[j \frac{k}{2z'} (x^2 + y^2)\right] \tag{4-38}$$

This corresponds directly with the expression for $h(x, y)$, when it is noted that if spatial variables (x, y) are used, one has to convolve $Q(x, y)$ with $U_t(x, y)$. Thus in summary, propagation is written either as[†] [30]

$$V(u, v) = V_t(u, v)Q(u, v) \tag{4-39}$$

or

$$U(x, y) = U_t(x, y) * Q_1(x, y) \tag{4-40}$$

where Eq. (4-40) is explicitly written out in Eq. (4-22).

In Chapter III it was shown that the effect of a lens on an incident wave was equivalent to multiplication with a spatial quadratic-phase term,

$$L_1(x, y) = \exp\left[-j \frac{k}{2f} (x^2 + y^2)\right] \tag{4-41}$$

The transmitted field is then

$$U_t(x, y) = U_i(x, y)L_1(x, y) \tag{4-42}$$

The transmitted field of a lens could also be expressed as

$$V_t(u, v) = V_i(u, v) * L(u, v) \tag{4-43}$$

where

$$L(u, v) = \exp\left[j \frac{f}{2k} (u^2 + v^2)\right] \tag{4-44}$$

[†] Note the suppression of the z'-dependent phase term.

Operational Techniques

The operations of propagation and lens modification have both been reduced to either multiplication by a quadratic-type term or convolution with a quadratic-type term. Having the results in this form suggests that an operational notation can possibly be developed that streamlines the work with optical system analysis. This notational method was developed by A. Vander Lugt [31].

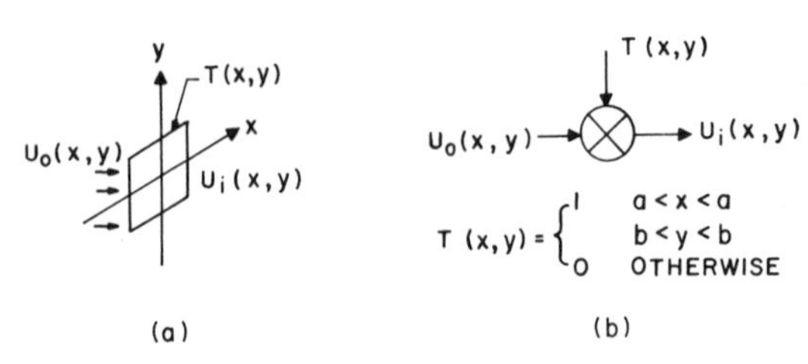

Fig. 4-2. Operational multiplier equivalents—physical space. (a) A transmitting aperture. (b) The multiplier equivalent.

Basically, the action of lenses and propagation will be considered as elements in a system. They will be denoted notationally as a multiplier, using the symbol ⊗, and as convolution, using a box ▭. If the spectrum variables are used, the multiplication element becomes a convolution element. Similarly, a convolution element over (x, y) becomes a multiplication over (u, v) space. Figure 4-2 contains an example of a bandlimiting aperture being represented by a multiplier element, that is, the product of the input field $u_i(x, y)$ and pupil function $t(x, y)$. In Fig. 4-3, the multiplier is being used in the dual representation, (u, v) space, to represent propagation. The spatial representation of a lens as a multiplier is shown in Fig. 4-4, and propagation in (x, y) variables is shown in Fig. 4-5.

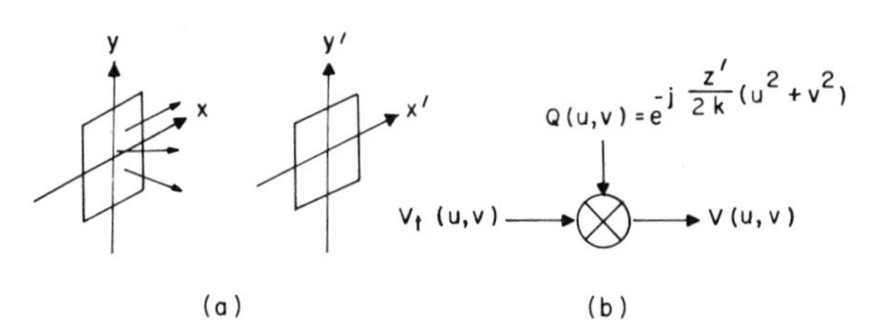

Fig. 4-3. Operational multiplier equivalents—transform space. (a) Propagation from one aperture (x, y) to another (x', y'). (b) The transform space multiplier equivalent.

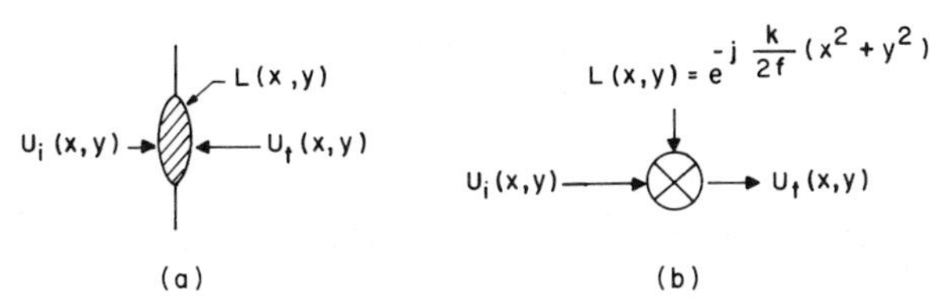

Fig. 4-4. Operational lens block diagram equivalent. (a) A physical space lens representation. (b) A physical space multiplier equivalent.

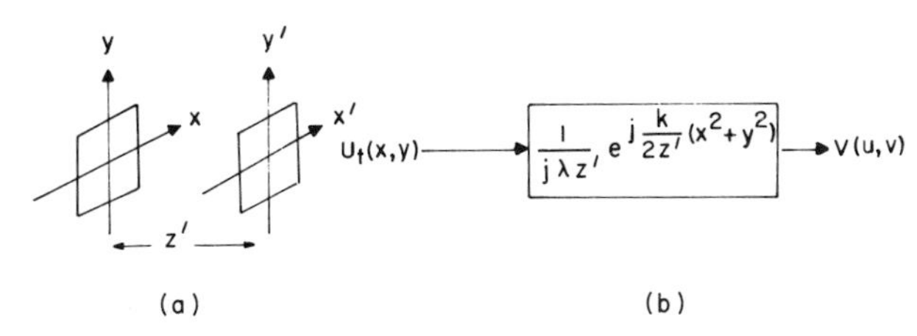

Fig. 4-5. Operational representation of propagation. (a) Propagation from one aperture to another. (b) A physical space convolutional equivalent.

Up to this point, only linear elements have been described. However, a complete optical system generally involves at least one nonlinear element in either the input or output of the system. Such an element is film. Just to complete the notation scheme, represent film as the product of a square-law detector and a νth law detector as shown in Fig. 4-6.

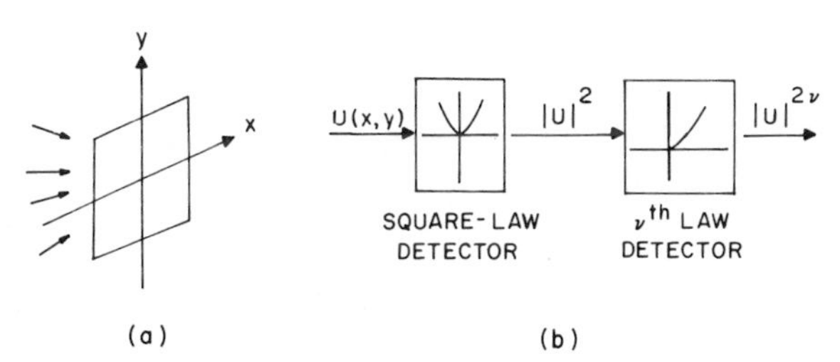

Fig. 4-6. Operational representation of film. (a) Exposing of film in an optical system. (b) A two-step multiplier equivalent.

The use of convolution or the box notation for propagation in physical space (x, y) allows an equivalence to be developed with circuit-analysis techniques. This equivalence suggests a representation of propagation by an impulse response, which mathematically is just the function† $Q_1(x, y)$ given in Eq. (4-38).

† For convenience the complex constant $1/j\lambda$ will be suppressed and will be carried like the linear phase terms.

A Canonical Function

Now a notation and the procedural use of this notation can be defined in terms of a limited algebra. To do so, the definition of a canonical function given by Vander Lugt can be given by

$$\psi(x, y; s) \equiv \exp\left[j \frac{ks}{2} (x^2 + y^2)\right] \tag{4-45}$$

where s is a reciprocal distance given by either

$$s = 1/z' \tag{4-46a}$$

or

$$s = 1/f \tag{4-46b}$$

The properties† of this ψ function are presented in Table 4-1.

Table 4-1

Functional Properties of the $\psi(x, y; s)$ Functions

$$\psi(x, y; s) = \bar{\psi}(x, y; -s) \tag{4-47a}$$

$$\psi(-x_1, -y; s) = \psi(x, y; s) \tag{4-47b}$$

$$\psi(x, y; s_1)\psi(x, y; s_2) = \psi(x, y; s_1 + s_2) \tag{4-48a}$$

$$\psi(x, y; s_1)\bar{\psi}(x, y; s_2) = \psi(x, y; s_1 - s_2) = \bar{\psi}(x, y; s_2 - s_1) \tag{4-48b}$$

$$\psi(x, y; s)\bar{\psi}(x, y; s) = 1 \tag{4-48c}$$

$$\psi(cx, cy; s) = \psi(x, y; c^2 s) \tag{4-49}$$

$$\bar{\psi}(x; s_1)\bar{\psi}(x, y; s_2) = \bar{\psi}(y, s_2)\bar{\psi}(x; s_1 + s_2) \tag{4-50a}$$

$$\bar{\psi}(y; s_1)\bar{\psi}(x, y; s_2) = \bar{\psi}(x, s_2)\bar{\psi}(y; s_1 + s_2) \tag{4-50b}$$

$$\psi(x - u, y - v; s) = \psi(x, y; s)\psi(u, v; s) \exp[-jks(ux + vy)] \tag{4-50c}$$

$$\psi(x, y; s) = \psi(x; s)\psi(y; s) \tag{4-50d}$$

$$\lim_{s \to 0} \bar{\psi}(x, y; s) = 1 \tag{4-51}$$

$$\lim_{s \to \infty} s\psi(x, y; s) = \delta(x, y) \tag{4-52}$$

$$\lim_{s \to 0} s\psi(x, y; s) = 0 \tag{4-53}$$

† The over bar on ψ means conjugate operation; i.e., $\bar{\psi}(x, y; s) = \exp[-j(ks/2)(x^2 + y^2)]$.

Equations (4-47) give the symmetry and asymmetry properties; Eqs. (4-48) give the rules for multiplication, including the identity element property, Eq. (4-48c). Equation (4-49) is the property of scale change, and Eqs. (4-50) give the factorization properties, including the decomposition of the convolution integral. It should be noted that the factors can be cylindrical lenses, spherical lenses, or combinations of each. Equation (4-51) expresses the limiting property of the function ψ as s goes to zero. Physically it says that infinite focal-length lenses do not affect the waves, except for phase delay. Equation (4-52) follows directly from our definition of the delta function used in Eq. (3-26). Equation (4-53) is similar to the initial value statement in system theory or Laplace transform theory. Note that Eqs. (4-52) and (4-53) form the analogues of the initial and final value theorems used in control theory and elsewhere. With this basic set of properties, first-order optical systems can be analyzed using skills from electrical linear systems analysis techniques. It will be a matter now of properly combining these operational elements.

Calculational Techniques for Imaging through a Lens

To see the utility of the notation, the imaging through a lens, conditions derived in Chapter III can be re-examined. The model will be that shown in Fig. 4-7. The first step is to write an equivalent block diagram in the spatial domain and then write the corresponding mathematical representation for each element. The multiplication block is straightforward. It is just the product of all elements into the multiplier using a common set of variables. The convolution blocks, which contain an integral of a product of an input function times a modified ψ function, are a little more complicated. Here one must identify the input and output variables and write the term in the block as a difference of input minus output variables of the

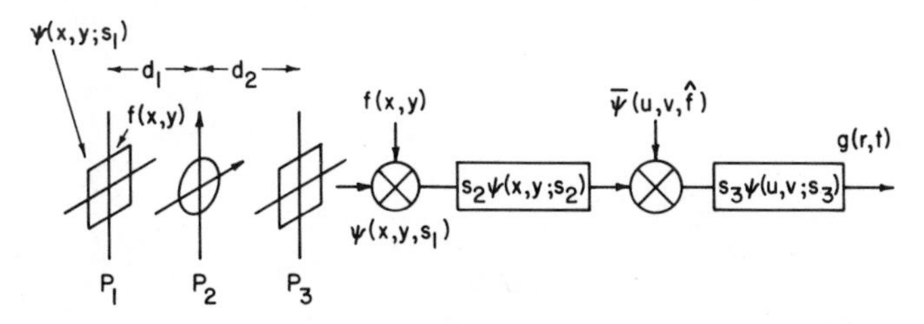

Fig. 4-7. A simple lens system and its block diagram equivalent.

argument. The other point to remember is the s_i multiplier term outside. It is also implicit in the convolution terms, to write it as an integral over the input variables.

Now using the block diagram and convolution blocks, the output can be written directly as

$$g(r, t) = s_2 s_3 \iint_{P_1} \iint_{P_2} \psi(x, y; s_1) f(x, y) \psi(x - u, y - v; s_2) \times \bar{\psi}(u, v; \hat{f}) \psi(u - r, v - t, s_3)\, dx\, dy\, du\, dv \tag{4-54}$$

where $\hat{f} = 1/f$. Using the convolution factorization property expressed in Eq. (4-50c), $g(r, t)$ becomes

$$\begin{aligned} g(r, t) = s_2 s_3 \iint_{P_1} \iint_{P_2} & \psi(x, y; s_1) f(x, y) \psi(x, y; s_2) \psi(u, v, s_2) \\ & \times \exp[-jks_2(ux + vy)] \bar{\psi}(u, v; \hat{f}) \psi(u, v, s_3) \psi(r, t; s_3) \\ & \times \exp[-jks_3(ur + vt)]\, dx\, dy\, du\, dv \end{aligned} \tag{4-55}$$

Equations (4-48a) and (4-48b) can now be used to combine terms, giving

$$\begin{aligned} g(r, t) = s_2 s_3 \psi(r, t; s_3) \iint_{P_1} \iint_{P_2} & \psi(x, y; s_1 + s_2) f(x, y) \psi(u, v; s_2 + s_3 - \hat{f}) \\ & \times \exp\{-jk[u(rs_3 + xs_2) + v(ts_3 + ys_2)]\}\, dx\, dy\, du\, dv \end{aligned} \tag{4-56}$$

It now becomes a matter of identifying which integrals can be done first. The best strategy seems to be to identify those that do not involve unknown functions. In this case it is the integration over (u, v). This integration can be done by using some of the work used in deriving Eq. (4-33). The reader can show this correspondence,[†] which leads to a general useful result:

$$k(u, v) \equiv \iint_{-\infty}^{\infty} \psi(x, y; s_1) \exp[-jks_2(ux + vy)]\, dx\, dy = \frac{j\lambda}{s_1} \bar{\psi}\left(u, v; \frac{s_2^2}{s_1}\right) \tag{4-57}$$

Applying Eq. (4-57) to Eq. (4-56) gives

$$\begin{aligned} g(r, t) = (j\lambda)\left(\frac{s_2 s_3}{s_2 + s_3 - \hat{f}}\right) \psi(r, t; s_3) \iint_{P_1} & \psi(x, y; s_1 + s_2) f(x, y) \\ & \times \bar{\psi}\left(x + \frac{s_3 r}{s_2}, y + \frac{s_3 t}{s_2}; \frac{s_2^2}{s_2 + s_3 - \hat{f}}\right) dx\, dy \end{aligned} \tag{4-58}$$

[†] See Problem 10 at the end of the chapter.

This result, using operational notation, is very much like the result obtained for Eq. (3-20) by direct application of the integral equation except that the constraint leading to the lens law has not been applied. It should be noted that this is a very compact notation lending itself to a large cascaded system analysis. In addition, the flexibility to examine the Fourier transforming condition as well as the simple imaging condition is easily provided. Further, one can do one-dimensional system analysis simply by appropriate factorization of this integrand using Eq. (4-50d).

Imaging Condition

To examine the simplest operation of imaging, define s_4 to be

$$s_4 = s_2^2/(s_2 + s_3 - \hat{f}) \tag{4-59}$$

and take the limit as

$$s_2 + s_3 - \hat{f} \to 0 \tag{4-60}$$

This limit forces s_4 to infinity, which, when Eq. (4-52) is used, reduces one term of the integrand to

$$\lim_{s_4 \to \infty} \left(\frac{s_3}{s_2}\right) s_4 \bar{\psi}\left(x + \frac{s_3 r}{s_2}, y + \frac{s_3 t}{s_2}; s_4\right) = \left(\frac{s_3}{s_2}\right) \delta\left(x + \frac{s_3 r}{s_2}, y + \frac{s_3 t}{s_2}\right) \tag{4-61}$$

If Eq. (4-61) is applied to the integrand of Eq. (4-58), the integration with the delta function simplifies the equation considerably, leaving

$$g(r, t) = \left(\frac{s_3}{s_2}\right) \psi\left(r, t; s_3 + (s_1 + s_2)\frac{s_3^2}{s_2^2}\right) f\left(-\frac{s_3 r}{s_2}, -\frac{s_3 t}{s_2}\right) \tag{4-62}$$

This result is simply the image of the function $f(x, y)$ in the output plane P_3.

Notice that $\psi(x, y; s_1)$ can be visualized as an exciting element and $f(x, y)$ considered as an input to the system. Further note that if $g(r, t)$ is recorded, $|g|^2$ is independent of the exciter since ψ is just a phase term which falls out under conjugation associated with intensity recording.

Equation (4-60) in the limit is just the *lens law*, Eq. (3-22), in reciprocal terms which is normally written as

$$1/d_1 + 1/d_2 = 1/f \tag{4-63}$$

where f is the focal length of the lens.

Fourier-Transform Condition

Following the same development of ideas as in Chapter III, the Fourier-transform condition can be derived. To do so, rearrange Eq. (4-58) so that Eq. (4-50c) can be used on the $\bar{\psi}$ of the integrand, giving

$$\bar{\psi}\left(x + \frac{s_3 r}{s_2}, y + \frac{s_3 t}{s_2}; \frac{s_2^2}{s_2 + s_3 - \hat{f}}\right)$$
$$= \psi\left(x + \frac{s_3 r}{s_2}, y + \frac{s_3 t}{s_2}; -\frac{s_2^2}{s_2 + s_3 - \hat{f}}\right)$$
$$= \psi\left(x, y; -\frac{s_2^2}{s_2 + s_3 - \hat{f}}\right)\psi\left(r, t, -\frac{s_3^2}{s_2 + s_3 - \hat{f}}\right)$$
$$\times \exp\left[-jk\left(\frac{s_2 s_3}{s_2 + s_3 - \hat{f}}\right)(rx + ty)\right] \tag{4-64}$$

Substituting Eq. (4-64) into Eq. (4-58), using the multiplicative formulas Eqs. (4-48), gives

$$g(r, t) = \psi\left(r, t; s_3 - \frac{s_3^2}{s_2 + s_3 - \hat{f}}\right) \iint_{P_1} \psi\left(x, y; s_1 + s_2 - \frac{s_2^2}{s_2 + s_3 - \hat{f}}\right)$$
$$\times f(x, y) \exp\left[-jk\left(\frac{s_2 s_3}{s_2 + s_3 - \hat{f}}\right)(rx + ty)\right] dx\, dy \tag{4-65}$$

The form of a Fourier transform is almost explicit, except that the quadratic-phase term $\psi[x, y; s_1 + s_2 - s_2^2/(s_2 + s_3 - \hat{f})]$ has to be reduced to unity. This then results in the appropriate transform kernel operating on $f(x, y)$.

If the basic properties of $\psi(\,)$ are examined, it is seen that Eq. (4-51) provides the proper vehicle to replace ψ by unity. Using this property, the condition for the Fourier transform is

$$s_1 + s_2 - s_2^2/(s_2 + s_3 - \hat{f}) = 0 \tag{4-66}$$

This result helps examine what different conditions on d_1, d_2, d_3, etc., lead to a Fourier transform. In the case of a plane wave s_1 goes to zero. Thus if a plane-wave input is used, Eq. (4-66) reduces to

$$s_2 - s_2^2/(s_2 + s_3 - \hat{f}) = 0 \tag{4-67}$$

which after some algebraic manipulation simplifies to

$$s_3 = \hat{f} \tag{4-68}$$

Physically, this corresponds to identifying the focal plane as the Fourier-transform plane. Further, from an intensity point of view there is no dependence on d_1 or d_2 in forming the Fourier-transform distribution in the focal plane. In this case then, Eq. (4-65) becomes

$$g(r, t) = \psi\left(r, t, \hat{f} - \frac{\hat{f}^2}{s_2}\right) \iint f(x, y) \exp[-jk\hat{f}(xr + yt)]\, dx\, dy \qquad (4\text{-}69)$$

However, the Fourier-transform relationship can be made an exact one for the fields by setting

$$s_2 = \hat{f} \qquad (4\text{-}70)$$

or physically placing the input function or mask in the front-focal plane of the lens. Notice that the ability to see what to do here was greatly facilitated by this operational notation. This result is obtained in a much easier fashion than was possible in obtaining Eq. (3-36).

Fourier Transform Using the Back Plane of the Lens

Now consider the possibility of using the back planes of a lens as an input region for the information mask and as a way of compressing the actual length of the optical systems. To do this, consider the system of Fig. 4-8 and its associated block diagram. As a matter of simplification, the first multiplier could be dropped through the use of Eq. (4-48b), reducing the input to

$$\psi(x, y; 0)\bar{\psi}(x, y, \hat{f}) = \bar{\psi}(x, y, \hat{f}) \qquad (4\text{-}71)$$

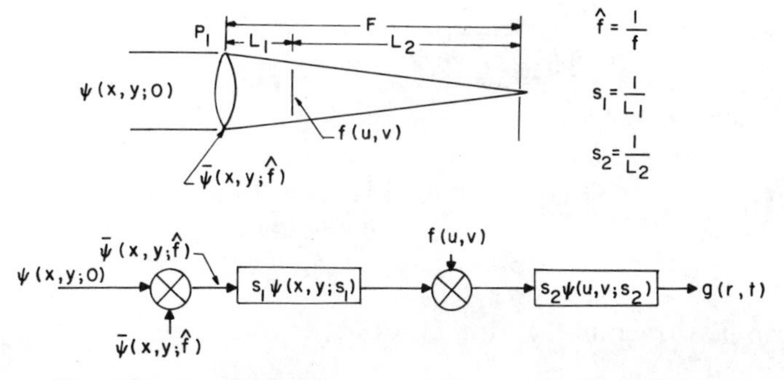

Fig. 4-8. A variable-scale transformer and its block diagram.

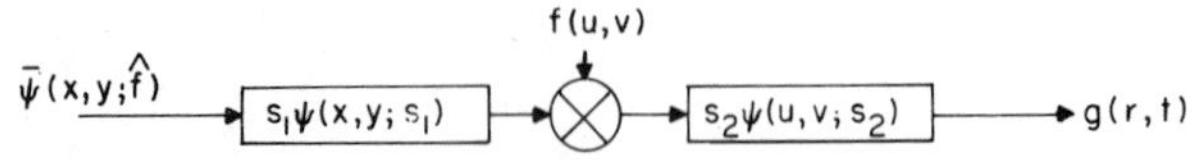

Fig. 4-9. A simplified form of variable-scale transformer block diagram equivalent.

leaving the block diagram of Fig. 4-9. Therefore, using this simplified block diagram, write

$$g(r,t) = s_1 s_2 \iint_{P_1} \iint_{P_2} \bar{\psi}(x, y; \hat{f})\psi(x-u, y-v; s_1) f(u,v) \times \psi(u-r, v-t, s_2)\, dx\, dy\, du\, dv \tag{4-72}$$

Using Eq. (4-50c), Eq. (4-72) can be simplified to

$$\begin{aligned} g(r,t) = s_1 s_2 \iint_{P_1} \iint_{P_2} & \bar{\psi}(x, y; \hat{f})\psi(x, y; s_1)\psi(u, v, s_1) \\ & \times \exp[-jks_1(ux+vy)] f(u,v)\psi(u,v,s_2)\psi(r,t,s_2) \\ & \times \exp[-jks_2(ur+vt)]\, dx\, dy\, du\, dv \end{aligned} \tag{4-73}$$

Further simplifications through the use of Eq. (4-48a) give

$$\begin{aligned} g(r,t) = s_1 s_2 \psi(r, t; s_2) \iint_{P_1} \iint_{P_2} & \psi(x, y; s_1 - \hat{f})\psi(u, v; s_1 + s_2) \\ & \times f(u,v) \exp[-jks_1(ux+vy)] \exp[-jks_2(ur+vt)]\, dx\, dy\, du\, dv \end{aligned} \tag{4-74}$$

The integration over (x, y) is done through the use of Eq. (4-57), giving

$$\begin{aligned} m(u,v) &\equiv \iint_{\infty}^{\infty} \psi(x, y; s_1 - \hat{f}) \exp[-jks_1(ux+vy)]\, dx\, dy \\ m(u,v) &= \frac{j\lambda}{s_1 - f}\, \bar{\psi}\left(u, v; \frac{s_1{}^2}{s_1 - \hat{f}}\right) \end{aligned} \tag{4-75}$$

and leaving for $g(r, t)$,

$$\begin{aligned} g(r,t) = \frac{s_1 s_2}{s_1 - \hat{f}} (j\lambda)\psi(r, t; s_2) \iint_{P_2} & \psi\left(u, v; s_1 + s_2 - \frac{s_1{}^2}{s_1 - \hat{f}}\right) \\ & \times f(u,v) \exp[-jks_2(ur+vt)]\, du\, dv \end{aligned} \tag{4-76}$$

This can be written in the form of a Fourier transform if

$$s_1 + s_2 - s_1{}^2/(s_1 - \hat{f}) = 0 \tag{4-77}$$

Equation (4-77) can be manipulated by proper transpositions to

$$\hat{f} = \frac{s_1 s_2}{s_1 + s_2} = \frac{1}{1/s_1 + 1/s_2} \tag{4-78}$$

However, the last half of this equation is just the constraint started with, namely,

$$f = L_1 + L_2 \tag{4-79}$$

Thus the condition for the Fourier transform is satisfied for all values of L_1 and L_2, leaving

$$g(r, t) = \left(\frac{s_1}{s_1 - \hat{f}}\right)^2 (j\lambda \hat{f}) \psi\left(r, t; \frac{s_1 \hat{f}}{s_1 - \hat{f}}\right) \times \iint_{P_2} f(u, v) \exp\left[-jk\left(\frac{s_1 \hat{f}}{s_1 - \hat{f}}\right)(ur + vt)\right] du\, dv \tag{4-80}$$

where Eq. (4-78) was used to obtain s_2 in terms of s_1 and $\hat{f}$ throughout the equation. From this equation, it is seen that scaling of variables or spectral scanning amplitude is a function of s_1, implying that the selection of s_1 affects the amount of illumination transmitted through the system and thus the overall intensity. This system can be used to search through various input scale sizes and is a natural for variable-scale transform studies, particularly when a peaking in spatial frequency is sought.

Cascaded Systems and System Operations

In general, many data-processing functions—correlation, autocorrelation, cross-correlation, convolution, integration, etc.—are representable by a general integral relation of the form

$$0(x, y) = \iint_{-\infty}^{\infty} f(u, v) h(x - u, y - v)\, du\, dv \tag{4-81}$$

This operation is a natural for the kind of systems being discussed. In some cases it is a repeated application of the Fourier-transform operation with appropriate modification of each transform. In the case of Eq. (4-81), it may be that $H(p, q)$ is to be multiplied by $F(p, q)$, where $H(p, q)$ is the Fourier transform corresponding to $h(x, y)$. The function $H(p, q)$ may represent, in reality, a mask, a variable thickness phase plate, propagation, or other combinations of processor elements.

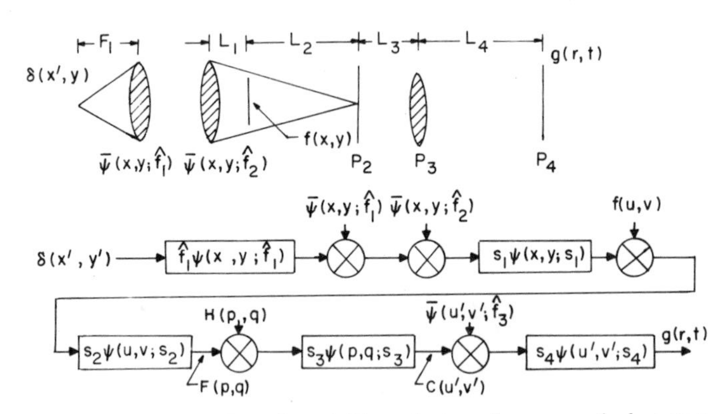

Fig. 4-10. A more complicated variable-scale transformer optical system and its block diagram.

To discuss this general data-processing question in detail, consider the system and its corresponding block diagram shown in Fig. 4-10.

The examination of this problem in considerable detail brings out some subtle implications of the analysis technique. First is the question as to whether sections of the system can be treated as separate elements. A similar question arises in electronic circuits with the loading of one circuit on another in cascaded circuits. Basically this relates to the separability of the optical processor into subcomponent systems. The problem arises when the phase error in various transform planes is not properly accounted for.

H(p,q) $\bar{\psi}(u', v'; \hat{f}_3)$

F(p,q) → ⊗ → $s_3\psi(p,q;s_3)$ → ⊗ → $s_4\psi(u',v',s_4)$ → g(r,t)

Fig. 4-11. A simplified form of the complicated variable-scale transformer block diagram equivalent.

In this problem of the variable-scale transformer (VST), the phase error needs to be carefully accounted for in the plane of $H(p, q)$. This will affect the determination of the conditions on $\hat{f}_3$ for inverse transforming of the product $H(p, q)$ and $F(p, q)$. This analysis begins with an examination of the significance of the delta-function excitation. The block diagram equivalent representation is shown in Fig. 4-11. It can be shown† through the use of the convolution integral that the equivalent output of the first lens, a focal distance from the point source, is a plane wave. This plane wave is then the input to the second lens and is represented by $\psi(x, y; 0)$. Thus

† See Problem 19 for details on this equivalence.

the problem reduces to what was done in the previous section. Using the results obtained in Eq. (4-80), the input at the multiplier associated with $H(p, q)$ can be written down. As a matter of simplification, abbreviate this as $F(p, q)$, giving

$$F(p, q) = \left(\frac{s_1}{s_1 - \hat{f}_2}\right)\hat{f}_2(j\lambda)\psi\left(p, q; \frac{s_1\hat{f}_2}{s_1 - \hat{f}_2}\right) \times \iint f(u, v) \exp\left[-jk\left(\frac{s_1\hat{f}_2}{s_1 - \hat{f}_2}\right)(pu + qv)\right] du\, dv \tag{4-82}$$

Using this result, Fig. 4-10 can be simplified to that shown in Fig. 4-11. Now the output expression for this cascaded system can be obtained quite easily using the techniques outlined earlier. Thus $g(r, t)$ becomes

$$g(r, t) = s_4 \iint du'\, dv'\, s_3 \iint dp\, dq\, F(p, q)H(p, q) \times \psi(p - u', q - v'; s_3)\bar{\psi}(u', v', \hat{f}_3)\psi(u' - r, v' - t, s_4) \tag{4-83}$$

This equation can be reduced through the factorization formulas. The objective in this factoring is to set the (u', v') integration apart. Thus $g(r, t)$ becomes

$$g(r, t) = s_3 s_4 \psi(r, t, s_4) \iint dp\, dq\, F(p, q)H(p, q)\psi(p, q; s_3) \times \iint du'\, dv'\, \psi(u', v'; s_3 + s_4 - \hat{f}_3) \times \exp\left\{-jks_3\left[u'\left(p + \frac{s_4}{s_3} r\right) + v'\left(q + \frac{s_4}{s_3} t\right)\right]\right\} \tag{4-84}$$

Using Eq. (4-57) gives

$$g(r, t) = \frac{s_3 s_4}{s_3 + s_4 - \hat{f}_3} (j\lambda)\psi(r, t; s_4) \iint dp\, dq\, F(p, q)H(p, q) \times \psi(p, q; s_3)\bar{\psi}\left(p + \frac{s_4}{s_3} r, q + \frac{s_4}{s_3} t; \frac{s_3{}^2}{s_3 + s_4 - \hat{f}_3}\right) \tag{4-85}$$

which, when separated, yields

$$g(r, t) = \frac{s_3 s_4}{s_3 + s_4 - \hat{f}_3} (j\lambda)\psi\left(r, t; s_4 - \frac{s_4{}^2}{s_3 + s_4 - \hat{f}_3}\right) \times \iint dp\, dq\, F(p, q)H(p, q)\psi\left(p, q; s_3 - \frac{s_3{}^2}{s_3 + s_4 - \hat{f}_3}\right) \times \exp\left[-jk\left(\frac{s_3 s_4}{s_3 + s_4 - \hat{f}_3}\right)(pr + qt)\right] \tag{4-86}$$

There is a great temptation at this point to say that the processor is complete and the appropriate inverse transform of $F(p, q)H(p, q)$ is obtained by forcing the $\psi(\,)$ function inside the integrand to unity. This method would yield a value of s_4 equal to $\hat{f}_3$. This decision, however, does not account for the phase-error term in Eq. (4-82). Thus Eq. (4-86) must be rewritten to account for the scale size changes in $F(p, q)$ and the total phase error by using a new function†

$$F'[(1 + s_2/s_1)p, (1 + s_2/s_1)q]$$

which shows directly the scaling factors. In this case $g(r, t)$ reduces to

$$\begin{aligned} g(r, t) = {} & \left(\frac{s_3 s_4}{s_3 + s_4 - \hat{f}_3}\right)\left(1 + \frac{s_2}{s_1}\right)^2 \hat{f}_2(j\lambda)^2 \psi\left(r, t; s_4 - \frac{s_4^2}{s_3 + s_4 - \hat{f}_3}\right) \\ & \times \iint dp\, dq\, H(p, q) \iint du\, dv\, f(u, v) \\ & \times \exp\left[-jk\left(1 + \frac{s_2}{s_1}\right)\hat{f}_2(up + vq)\right] \\ & \times \psi\left(p, q, \hat{f}_2 + \frac{s_2 \hat{f}_2}{s_1} + s_3 - \frac{s_3^2}{s_3 + s_4 - \hat{f}_3}\right) \\ & \times \exp\left[-jk\left(\frac{s_3 s_4}{s_3 + s_4 - \hat{f}_3}\right)(pr + qt)\right] \end{aligned} \tag{4-87}$$

Now the condition for the Fourier transform becomes

$$\hat{f}_2 + \frac{s_2}{s_1}\hat{f}_2 + s_3 - \frac{s_3^2}{s_3 + s_4 - \hat{f}_3} = 0 \tag{4-88}$$

which can be written as

$$s_2 + \frac{s_3(s_4 - \hat{f}_3)}{s_3 + s_4 - \hat{f}_3} = 0 \tag{4-89}$$

or, if L_2, L_3, L_4, and f_3 are substituted, becomes

$$1/f_3 = 1/L_4 + 1/(L_2 + L_3) \tag{4-90}$$

Thus an imaging condition on the output system is obtained. The output

† Note that use has been made of the condition that

$$\frac{s_1 \hat{f}_2}{s_1 - \hat{f}_2} = \left(\frac{1/\hat{f}_2}{1/\hat{f}_2 - 1/s_1}\right)\hat{f}_2$$

which, for the condition in Eq. (4-78), reduces to $(1 + s_2/s_1)\hat{f}_2$.

lens is then placed such that it images the input mask to the output plane. That is, P_1 is imaged through $H(p, q)$ to P_4.

If it is desired that the function $g(r, t)$ is to represent the inverse Fourier transform, then $\psi[r, t; s_4 - (s_4^2/s_3 + s_4 - \hat{f}_3)]$ must reduce to unity, which occurs for $s_3 = \hat{f}_3$. However, notice that this condition is incompatible with Eq. (4-90). Thus it is seen that the phase-error term must be retained. Generally, however, only the intensity of the output is recorded; thus this phase error will not affect the recorded output distribution.

The only other constraint is that of the total length $(L_1 + L_2)$ derived in the variable-scale transform condition of $f_2 = L_1 + L_2$. Thus $g(r, t)$ can be written as

$$g(r, t) = \left(\frac{s_2}{\hat{f}_3 - s_4}\right)\left(1 + \frac{s_2}{s_1}\right)^2 \hat{f}_2(j\lambda)^2 \psi\left[r, t; \frac{s_4(s_3 - \hat{f}_3)}{s_3 + s_4 - \hat{f}_3}\right]$$
$$\times \iint dp\, dq\, H(p, q) F'\left[\left(1 + \frac{s_2}{s_1}\right)p, \left(1 + \frac{s_2}{s_1}\right)q\right]$$
$$\times \exp\left[-jk\left(\frac{s_2}{\hat{f}_3 - s_4}\right)(pr + qt)\right] \tag{4-91}$$

At this point it becomes apparent that one could make a change of variables over (p, q) incorporating the factor $(1 + s_2/s_1)$ and represent the integral as an inverse transform over a variable scale filter. In either case, the result shows that a composite system can be obtained which will search through various spatial frequency sizes and enable a general filtering operation on arbitrary scale size input masks. This has the advantage that input data and system alignment are not so dependent on exact scale sizes, and general filters can be produced and used.

Multichannel One-Dimensional Systems

Multichannel one-dimensional systems can also be analyzed and synthesized by the operational notation through the use of Eqs. (4-50a) to (4-50c). A multichannel spectrum analyzer is shown in Fig. 4-12 with its corresponding block diagram. The equation for the output function $g(r, t)$ after using Eqs. (4-50b) and (4-50c) is

$$g(r, t) = s_3 s_2 \psi(r, t; s_3) \iint_{P_1} \iint_{P_2} f(x, y)\psi(x, y; s_1 + s_2)\psi(v; s_2 - \hat{f}_1 + s_3 - \hat{f}_2)$$
$$\times \psi(u; s_2 - \hat{f}_2 + s_3) \exp[-jks_2(ux + vy)]$$
$$\times \exp[-jks_3(ur + vt)]\, dx\, dy\, du\, dv \tag{4-92}$$

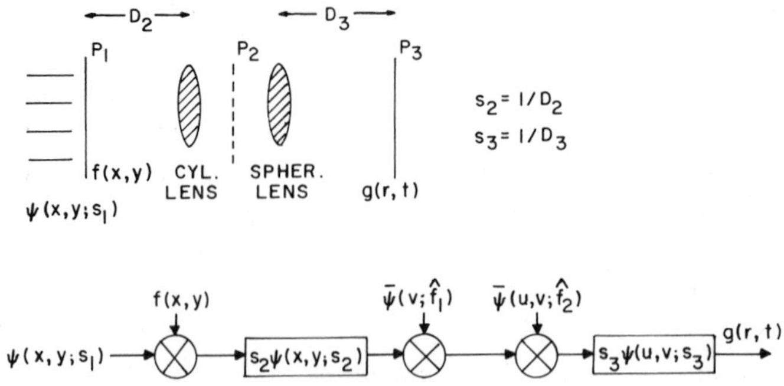

Fig. 4-12. A one-dimensional multichannel system and block diagram equivalent.

The integration over (u, v) can be done by procedures similar to those used in obtaining Eq. (4-57). Considering only the (u, v) integral gives

$$\iint_{P_2} du\, dv\, \psi(v; s_2 - \hat{f}_1 + s_3 - \hat{f}_2)\psi(u; s_2 - \hat{f}_2 + s_3)$$
$$\times \exp\left[-jks_2u\left(x + \frac{s_3}{s_2} r\right)\right] \exp\left[-jks_2v\left(y + \frac{s_3}{s_2} t\right)\right]$$
$$= (j\lambda) \frac{1}{[(s_2 - \hat{f}_1 + s_3 - \hat{f}_2)(s_2 - \hat{f}_2 + s_3)]^{1/2}}$$
$$\times \bar{\psi}\left(x + \frac{s_3}{s_2} r; \frac{{s_2}^2}{s_2 - \hat{f}_2 + s_3}\right) \bar{\psi}\left(y + \frac{s_3}{s_2} t; \frac{{s_2}^2}{s_2 - \hat{f}_1 + s_3 - \hat{f}_2}\right) \tag{4-93}$$

which, when substituted back in Eq. (4-92), leaves

$$g(r, t) = \frac{(j\lambda)\psi(r, t; s_3)s_2s_3}{[(s_2 - \hat{f}_1 + s_3 - \hat{f}_2)(s_2 - \hat{f}_2 + s_3)]^{1/2}} \iint f(x, y)\psi(x, y; s_1 + s_2)$$
$$\times \bar{\psi}\left(x + \frac{s_3 r}{s_2}; \frac{{s_2}^2}{s_2 - \hat{f}_2 + s_3}\right)$$
$$\times \bar{\psi}\left(y + \frac{s_3 t}{s_2}; \frac{{s_2}^2}{s_2 - \hat{f}_2 + s_3 - \hat{f}_1}\right) dx\, dy \tag{4-94}$$

With Eq. (4-94) in this form, a type of single-channel processing can be done. For example, it can be elected to do a Fourier transform in x and direct imaging in y. To have imaging in y, a delta function over y must be formed through the use of Eq. (4-52).[†] Thus the following limit must be taken

$$\lim_{s_2-\hat{f}_1+s_3-\hat{f}_2\to 0} \frac{s_2^{\,2}}{(s_2-\hat{f}_1+s_3-\hat{f}_2)^{1/2}}\, \psi\!\left(y+\frac{s_3 t}{s_2};\, \frac{s_2^{\,2}}{s_2-\hat{f}_2+s_3-\hat{f}_1}\right)$$

$$\rightarrow \delta\!\left(y+\frac{s_3 t}{s_2}\right) \tag{4-95}$$

which implies that

$$s_2-\hat{f}_1+s_3-\hat{f}_2=0 \tag{4-96}$$

or

$$\frac{1}{D_2}+\frac{1}{D_3}=\frac{f_2+f_1}{f_1 f_2} \tag{4-97}$$

Thus this condition, Eq. (4-97), implies that the two lenses act as thin lenses in contact and is a statement of the imaging condition in the y direction.

Applying Eqs. (4-95) and (4-97) to Eq. (4-94) and integrating with the delta function gives

$$g(r,t)=\frac{(j\lambda)s_3}{\sqrt{\hat{f}_1}}\,\psi(r;s_3)\psi\!\left[t;\, s_3+\left(\frac{s_3}{s_2}\right)(s_1+s_2)\right]\int f\!\left(x_1-\frac{s_3}{s_2}t\right)$$

$$\times\, \psi\!\left(x;\, s_1+s_2-\frac{s_2^{\,2}}{s_2+s_3-\hat{f}_2}\right)$$

$$\times\, \exp\!\left[-jk\left(\frac{s_2 s_3}{s_2+s_3-\hat{f}_2}\right)xr\right] dx \tag{4-98}$$

For this to be in the form of a Fourier transform, the ψ function in the integrand must go to unity. This can be accomplished by having

$$s_1+s_2-\frac{s_2^{\,2}}{s_2+s_2-\hat{f}_2}=0 \tag{4-99}$$

This condition is similar to that derived in Eq. (4-66). If $s_1=0$, as in the case of a plane wave, then Eq. (4-99) is satisfied for all values of s_2 and further $s_3=\hat{f}_2$. However, if s_3 equals $\hat{f}_2$, then from Eq. (4-96) s_2 equals $\hat{f}_1$, forcing $f(x,y)$ to be one focal length away from the cylindrical lens.

[†] Note that Eq. (4-52) arises from taking the limiting form on a gaussian function which is like $(1/\sqrt{\sigma})e^{-x^2/\sigma}$, yielding $\lim_{s\to\infty}\sqrt{s/\pi}\,\psi(x;s)=\delta(x)$.

With these conditions $g(r, t)$ becomes

$$g(r, t) = -(j\lambda)\frac{s_3}{\sqrt{\hat{f}_1}}\,\psi(r; \hat{f}_2)\psi\left(t; \hat{f}_1 + \frac{\hat{f}_2{}^2}{\hat{f}_1}\right)$$
$$\times \int f\left(x_1 - \frac{\hat{f}_2}{\hat{f}_1}t\right)\exp[-jk\hat{f}_2 xr]\,dx \tag{4-100}$$

For unit magnification, $\hat{f}_1 = \hat{f}_2$, this reduces $g(r, t)$ to

$$g(r, t) = (j\lambda)\sqrt{\hat{f}}\;\psi(r; \hat{f})\psi(t; 2\hat{f})\int f(x_1 - t)\exp[-jk\hat{f}xr]\,dx \tag{4-101}$$

Thus a multichannel one-dimensional system has been synthesized. The implication is that planes P_1 and P_3 are regarded as ends of a black box. Note that this solution was developed without a priori knowledge of the optics within. However, one must always keep in mind that to use this optical system as an element in a larger optical system, the output phase error, which is asymmetrical, must be accounted for. In the simplest form this would amount to placing a cylindrical lens in the output plane. Total correction could be obtained with a spherical and a cylindrical lens in the output plane.

Bandlimiting Nature of Physical Systems

Throughout this chapter, it has been assumed that lenses were infinite in extent in the transverse direction. It is known that this is not physically true, and in the last chapter, this was accounted for by defining a pupil function. This problem is explicitly associated with Eq. (4-57). This equation had infinite limits, and the integral was performed using known definite integrals.

If Eq. (4-57) has finite limits, then the integral takes the form of Fresnel integrals similar to those used in Chapter III with Eq. (3-20). Equation (3-24) with sinc functions results when rectangular coordinate systems are considered. If circular lenses are used, then a cylindrical coordinate system has to be considered as done in Chapter II when studying circular apertures. In that case one would obtain a result for $k(\,)$ which contains a Bessel function of the first kind [16]; namely,

$$k(\varrho) = \pi A^2\left(\frac{2J_1(kA\hat{f}_2\varrho)}{kA\hat{f}_2}\right) \tag{4-102}$$

where A is the radius of the lens, $\hat{f}_2$ is the inverse focal length, and ϱ is the radial coordinate direction. The output of the lens is then obtained by convolving $k(\varrho)$ with $\psi(\varrho; s_1 + s_2)f(\varrho, \theta)$, where s_1 is associated with the curvature of the input wave, s_2 is associated with the lens placement, and $f(\varrho, \theta)$ is the input plane information function. Note that the overall system performance is governed by A, and in the limit of large A, the results reduce to the previous case.

In the derivation of the Fourier-transform relationship, however, all the values of the system dimensions were not specified. It will be shown that finite lens apertures place restrictions on the amount of data that can be processed, and on the positions of the input data and the Fourier-transform plane relative to the lenses. These restrictions are often replaced by more stringent ones [32] caused by lens aberrations, including curvature of wave front, coma, etc., but they do give some indication of the optimum results that one can expect.

As mentioned in Chapter III, the concept of transfer function and Fourier-transform representation was dependent on space invariance. From a physical point of view, the transform operation must operate uniformly over all the input-data plane. To give a heuristic picture of how space invariance works [31], consider the setup shown in Fig. 4-13.

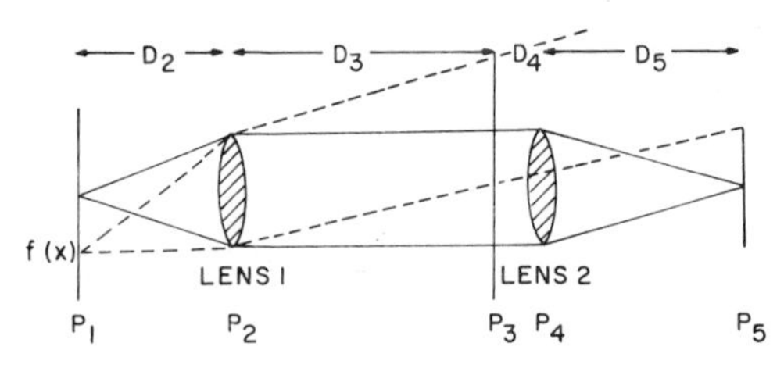

Fig. 4-13. A physical notion for optical bandlimiting in a simple two-lens optical system.

In this one-dimensional scheme, it can be seen how the problems arise, and gain the insight for real-life synthesization. A function $f(x)$ is placed in plane P_1, a distance D_2 from a lens having focal length f_1 and radius A_1. A lens having focal length f_2 and radius A_2 is placed near plane P_3. Suppose $f(x)$ consists of two point sources of light, one on axis and the other off axis as shown. For convenience, let $D_2 = D_3 = f_1$. The point source located on the optical axis will produce a plane wave in plane P_3, but only part of this wave will enter the second lens. Since P_5 is illuminated differently for each source, the input data was not invariant to its location with respect to the axis. Therefore, space invariance is not obtained, and there is no inherent advantage to having $D_2 = f_1$.

As mentioned before, space invariance also entered the picture in the Fourier-transform operation. In the one-dimensional picture, Eq. (4-57) becomes, for a lens of width $2A$,

$$k(u) = \frac{\sqrt{j}}{\sqrt{\hat{f}_1}}\, \bar{\psi}\left(u; \frac{\hat{f}_2^{\,2}}{\hat{f}_1}\right) k_1(u) \tag{4-103}$$

where

$$k_1(u) = \int_{-(k/\pi \hat{f}_1)^{1/2}(\hat{f}_1 A + \hat{f}_2 u)}^{(k/\pi \hat{f}_1)^{1/2}(\hat{f}_1 A - \hat{f}_2 u)} \exp\left[j \frac{\pi}{2} t^2\right] dt \tag{4-104}$$

The final result is similar, except that multiplying function is now $k_1(u)$, a Fresnel integral.

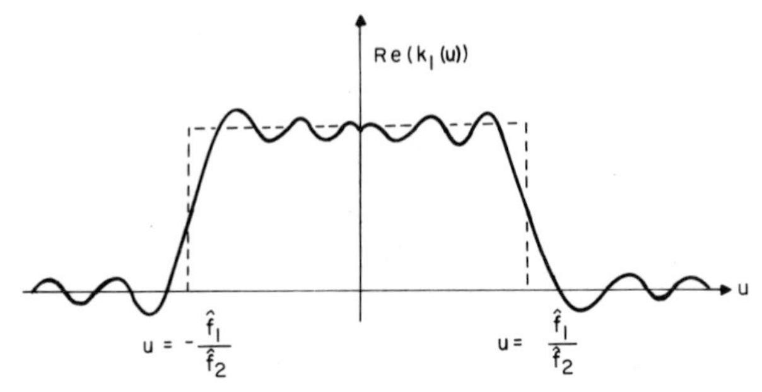

Fig. 4-14. A representation of $\mathrm{Re}[k_1(u)]$ and its approximate equivalent shown by the dotted lines.

The function $k_1(u)$ can be represented by the curve shown in Fig. 4-14. The wiggly line is associated with the familiar Gibbs phenomena and is caused by the oscillations that the Fresnel integral goes through for arguments near $|u| = (\hat{f}_1/\hat{f}_2)A$. Also shown in the Fig. 4-14 is a dotted square pulse which will be the approximation to $k_1(u)$. Some minor oscillations are lost, but the important bandlimiting characteristics are preserved. The approximate representation of $k_1(u)$ can be written[†]

$$k_1(u) = U(\hat{f}_2 u + \hat{f}_1 A) - U(\hat{f}_2 u - \hat{f}_1 A) \tag{4-105}$$

[†] The unit step function is unity for positive argument and zero for negative argument, i.e.,

$$U(t - t_0) = \begin{cases} 1 & t \geq t_0 \\ 0 & t < t_0 \end{cases}$$

If this $k_1(u)$ is applied to the general imaging work, Eq. (4-58) becomes

$$g(r, t) = -(j\lambda)\left(\frac{s_2 s_3}{s_2 + s_3 - \hat{f}}\right)\psi(r, t; s_3) \iint_{P_1} \psi(x, y; s_1 + s_2) f(x, y)$$

$$\times\ \bar{\psi}\left(x + \frac{s_3 r}{s_2}, y + \frac{s_3 t}{s_2}; \frac{s_2{}^2}{s_2 - \hat{f} + s_3}\right)$$

$$\times \left\{U\left[s_2\left(x + \frac{s_3}{s_2} r\right) + (s_2 - \hat{f} + s_3)A_x\right]\right.$$

$$\left. - U\left[s_2\left(x + \frac{s_3 r}{s_2}\right) - (s_2 - \hat{f} + s_3)A_x\right]\right\}$$

$$\times \left\{U\left[s_2\left(y + \frac{s_3}{s_2} t\right) + (s_2 - \hat{f} + s_3)A_y\right]\right.$$

$$\left. - U\left[s_2\left(y + \frac{s_3 t}{s_2}\right) - (s_2 - \hat{f} + s_3)A_y\right]\right\} dx\, dy \tag{4-106}$$

The effect of this $k_1(u, v)$ on the Fourier-transform operation for a lens system can be seen by examining one dimension for $s_1 = 0$ (plane-wave input). In this case $g(r)$ is obtained

$$g(r) = \psi\left(r; \hat{f} - \frac{\hat{f}_2}{s_2}\right) \int_{P_1} f(x) \exp(-jk\hat{f}xr)\left[U\left(x + \frac{s_3 r}{s_2} - A\right)\right.$$

$$\left. - U\left(x + \frac{s_3 r}{s_2} + A\right)\right] dx \tag{4-107}$$

which can be written as

$$g(r) = \psi\left(r; \hat{f} - \frac{\hat{f}_2}{s_3}\right) \int_{-(s_3 r/s_2)-A}^{(s_3 r/s_2)+A} f(x) \exp[-jk\hat{f}xr]\, dx \tag{4-108}$$

This result is very much like the previous work in finite apertures and establishes that the Fourier-transform operation is very much affected by the lens size A.

This result can be translated into meaningful constraints on the input-data format. First of all, note that for space invariance

$$g(r) \neq 0 \qquad \text{for} \quad |r| \leq A_2 \tag{4-109}$$

Thus, if this is applied to the boundary of the square pulse, s_2 and D_2 become

$$s_2 > \frac{s_3 A_2}{A_1 - x_0} \tag{4-110}$$

or

$$D_2 < \frac{D_3(A_1 - x_0)}{A_2} \tag{4-111}$$

where x_0 is associated with the edge of the input-data format, and A_1 is the input lens size. The length of the input data then becomes $2x_0$, and Eq. (4-111) provides a basis for system design, lens placement, and data placement.

The spatial frequency content of $f(x, y)$ determines A_2, and Eq. (4-111) in turn specifies the location of $f(x, y)$. Note that as $x_0 \to A_1$ (the size of the input data approaches the lens size), D_2 tends to zero, meaning that the input data must be placed very close to the lens in order that the band-limiting nature of the system does not take over. This result implies an inherent advantage of using the front planes of the lens as input regions as described in the section on variable-scale transforms. The two-dimensional extension for these results are obvious and are left to the reader as an exercise.

Problems

1. Assume that the field of an aperture is represented by

$$U(x, y) = \begin{cases} A & -b < y < b, \quad -a < x < a \\ 0 & \text{otherwise} \end{cases}$$

What is the spectrum of this aperture? What is the field at a point $z > 0$?

2. Find the spectrum of a plane wave represented by

$$U(x, y) = A \exp[i(k_x x + k_y y)]$$

Hint: Remember that one definition of a delta function is [1-2]

$$\int_{-\infty}^{\infty} e^{jax}\, dx = \delta(a)$$

3. Consider that $U(x, y, z)$ is a plane wave of the following form:

$$U(x, y, z) = \exp[i(k_x x + k_y y + k_z z)]$$

Find the aperture field and transform of the aperture field for the aperture defined by Eqs. (4-16) and (4-17) by both methods as given by Eqs. (4-18) and (4-15).

4. Given the series $f(s, \tau)$ defined by

$$f(s, \tau) = 1 + e^{-s\tau} + e^{-2s\tau} + e^{-3s\tau} + \cdots + e^{-ns\tau} + \cdots$$

write a closed form for this series.

5. Using the results of Problems 3 and 4, write the spatial Fourier transform of a series of rectangular gratings as shown in Fig. 4-1. Assume a plane-wave illumination.

6. Show that Eq. (4-32) is correct.

7. Show that the equation for $Q_1(x, y)$ does indeed correspond to this $Q(u, v)$.

8. Prove Eq. (4-57).

9. Using techniques similar to the derivation of Eq. (4-57), show that Eq. (4-75) is correct.

10. Prove the properties given by Eqs. (4-28) through (4-53).

11. Show that Eq. (4-78) follows from Eq. (4-77).

12. Show that the amplitude coefficient in Eq. (4-80) is the same as that in Eq. (4-76).

13. Show that Eq. (4-86) results from Eqs. (4-84) and (4-57).

14. Show how Eq. (4-89) results from Eq. (4-88).

15. Show how Eq. (4-90) results from Eq. (4-89).

16. Derive the scale factor $(1 + s_2/s_1)$ used in Eq. (4-91).

17. Find $g(r, t)$ for the system shown in Fig. 4-15. What are the conditions on L_1, L_2, L_3, L_4, and L_5 in order that $g(r, t)$ be the Fourier transform of $H(p, q)F(p, q)$? How can scale size changes be obtained? Be sure to draw the appropriate block diagram.

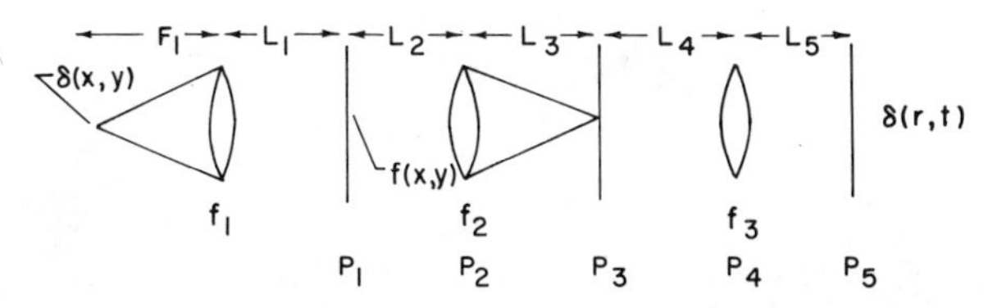

Fig. 4-15. A complete optical processor system in physical space.

18. Using the operational notation, show that two lenses cascaded can be represented by an equivalent lens of focal length $1/f_3 = 1/f_1 + 1/f_2$.

19. Given a lens excited by a point source $\delta(x', y')$ as shown in Fig. 4-16, find the output function $\psi(x, y; s)$ of the lens for $f = L_1$.

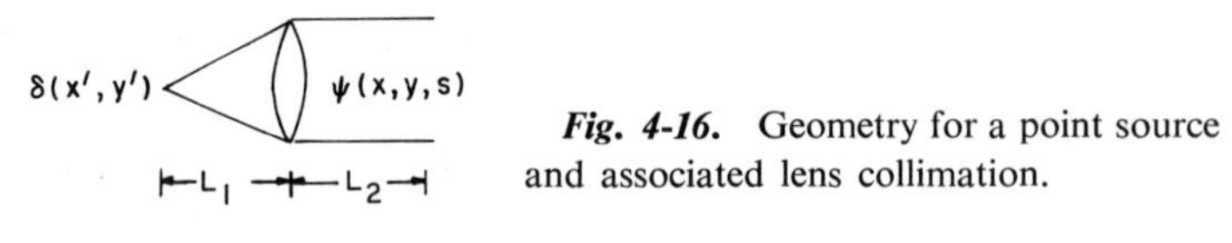

Fig. 4-16. Geometry for a point source and associated lens collimation.

20. Show how Eqs. (4-103) and (4-104) are obtained.

21. Derive an equivalent bandlimiting condition for two dimensions as that given in Eq. (4-111) for one dimension.

Chapter V

APPLICATIONS OF OPTICAL FILTERING TO DATA PROCESSING

Introduction

The preceding four chapters have set the foundation for the use of optical systems in data processing. The development started from first principles embodied in Maxwell's equations to the use of the operational notation to handle large cascaded systems. It remains then to examine applications of this theory.

To cover the whole area of applications is too much for one book, and there are many good treatises on applications [13, 22–24, 32, 33]. Thus it will suffice to cover some elementary examples as illustrative of the kind of processing that can be done in such systems, and then show some current examples in new areas of application, such as the biomedical problem area.

Band-Pass Filters

The simplest filtering that one can examine is the basic spatial band-pass type. It may be either low pass, band pass, or high pass. In addition, it may have low, band, or high pass in one direction and low, band, or high pass in the other direction.

The first demonstration that the spatial domain could indeed be modified was the Abbe–Porter experiment [34, 35]. As an example, consider the case of a two crossed grating shown in Fig. 5-1. The basic optical system with a grating input is shown in Fig. 5-1(a). Figure 5-1(b) shows the output in the focal plane.

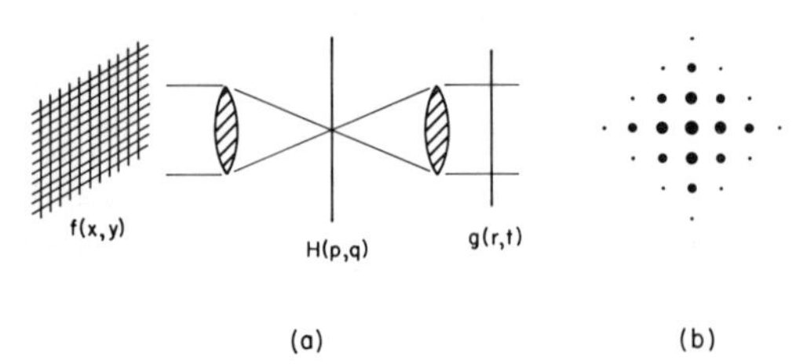

Fig. 5-1. Crossed gratings and their optical transforms. (a) Optical system with grating input. (b) Focal or Fourier plane output.

First, let $H(p, q)$ be a slit such that in the first case only the low frequencies in the x direction and all the frequencies in the y direction along the y axis will be passed, as shown in Fig. 5-2. In this case, all the information regarding the vertical grating is suppressed, and the information regarding the horizontal grating is passed. This then allows, upon reconstruction with a lens, an image of the horizontal grating to be created. One would refer to this as a "x low-pass filter." From symmetry arguments, a "y low-pass" filter would give a reimaged vertical grating and suppress the horizontal grating. Thus a very simple means of separating some kinds of images is obtained. In essence, this would be a form of demodulation if the gratings were spatial carriers.

If $H(p, q)$ were a spot on the optical axis, then all the highs but no dc component would be passed. This change in average level would cause a lot of scene reversal. Light areas would become dark and dark areas would become light, and the picture would appear to be a negative of the original scene. This can be illustrated by considering what happens as the average value is removed from Fig. 5-3. The amplitude A is plotted in Fig. 5-3(a) and (b), where the intensity I is plotted in Fig. 5-3(c). Notice that the net effect in Fig. 5-3(c) is a loss of detail and an apparent enhancement of the edges. In some processing problems edge enhancement is a desirable result. In general, however, the original scene information is reduced due to the loss of the dc or average component.

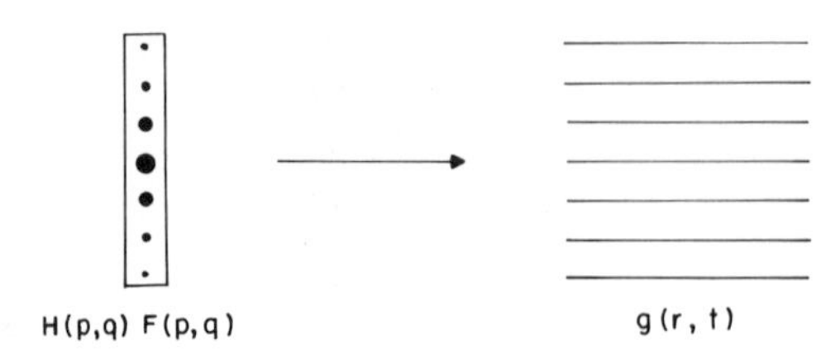

Fig. 5-2. Effect of simple band-pass filter on a complex grating system. Reconstruction of a simple horizontal grating.

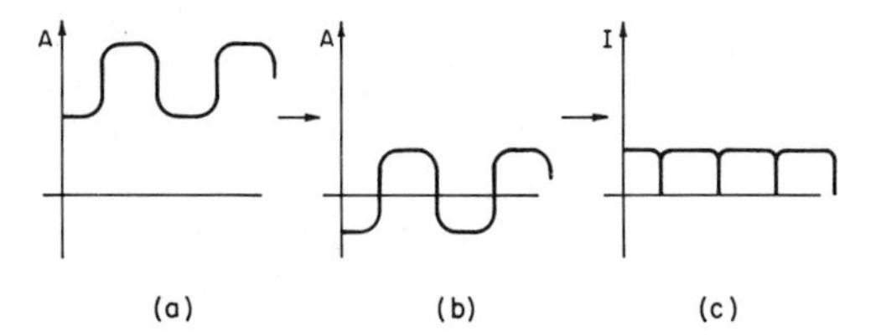

Fig. 5-3. Effect of level change. (a) Amplitude level on original input versus position. (b) Change of level by removal of average value. (c) Intensity distribution of (b).

Tiger-in-the-Cage

As a final example of this type, consider the following bit of magic. If a picture of a tiger in a cage is used, the cage can be made to disappear by suppression of the information about the cage bars, as shown in Fig. 5-4. The filter to do this would be a one-dimensional stop-band filter, which would be oriented in a direction perpendicular to the bars. A typical example would be like the one shown in Fig. 5-5.

The tiger-in-the-cage principle can be applied to many experiments where specific kinds of periodic data need to be suppressed. An example is the processing of the strip photos taken of the moon and Mars. The pictures are composed of strips and the lines associated with the strip edge are removed by suppressing the periodic grating associated with the lines.

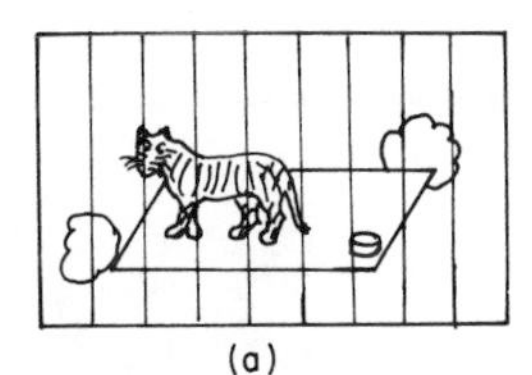

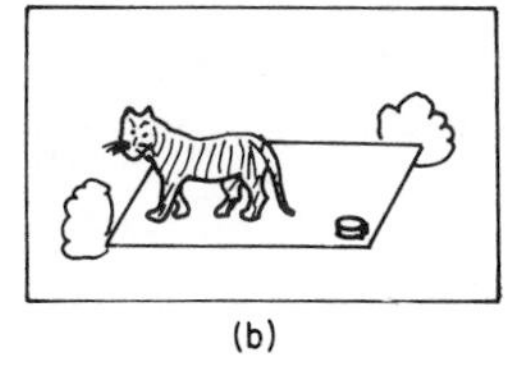

Fig. 5-4. Simple filtering for removal of unwanted information. Let the tiger out by removal of the vertical grating information in the focal plane.

Fig. 5-5. A simple angular band-stop filter that can remove information along a particular angle in focal plane. Filter used in letting tiger out of the cage figure.

Edge Sharpening-Contrast Improvement

Returning to the one-dimensional picture, consider what would happen if the average value or dc value was changed in the focal plane. The signal-to-noise ratio can be increased by partial changes in the dc value. This is demonstrated in Fig. 5-6. Note that as the average value is decreased, the apparent signal-to-noise ratio increases. Note that if this process is carried too far, the average can be brought down to the point that fold-over results, and image reversal would occur. This was demonstrated in the previous section. High contrast is being associated with maximum signal-to-noise ratio. This simple illustration shows that marked improvement of the contrast of a picture can be obtained by controlled changes in the average value or dc term in the focal plane.

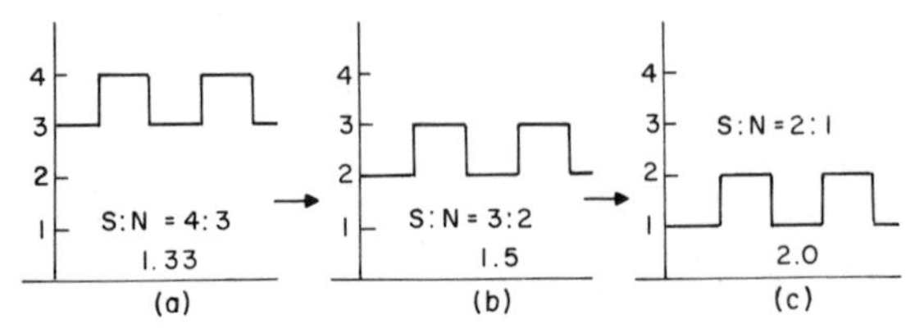

Fig. 5-6. Signal-to-noise improvement with average level changes. (a) Original level versus position. (b) Some level change by removal of some average value in focal plane. (c) Additional level change.

Another way to view this operation is the enhancement of the high-frequency components and deemphasis of the lows. This has the effect of sharpening up edges and bringing out fine detail. Numerous examples of this are provided in the literature [23]. There are limitations to this scheme, and the most common output effect is a ringing effect on the edge of objects, the addition of edges where none were before.

Continuously Varying Masks

Up to this point only examples of discrete spatial-frequency filters have been considered. Moreover, primary focus has been on binary filters, that is, filters that either transmit or block the light. However, one can think in terms of continuously modifying the focal-plane distribution. Thus examples of continuously varying marks will be considered.

a. *Differentiation*

Another example of edge sharpening comes from a type of high-frequency weighting which is frequency dependent. One might heuristically ask what the effect of linear weighting might be. That is, suppose a filter is constructed which weights the higher-frequency components in such a way that the amount of transmission through the mask is linearly related to frequency. For example [36], let

$$H(p, q) = jp \tag{5-1}$$

From previous comments under the section on edge sharpening, it is obvious that the edges will be markedly enhanced and very possibly only edge information will appear.

Examination of the inverse transform of $F(p, q)H(p, q)$ reveals what happens. Consider, the inverse transform of $F(p, q)$ after transmission through a filter $H(p, q)$ represented by

$$g(r, t) = \left(\frac{1}{2\pi}\right)^2 \iint_{-\infty}^{\infty} jpF(p, q) \exp[j(pr + qt)]\, dp\, dq \tag{5-2}$$

If the properties of Laplace-transform theory are considered, one can note that

$$jp \exp[j(pr + qt)] \,(=)\, \frac{\partial}{\partial r} \exp[j(pr + qt)] \tag{5-3}$$

Thus there is an equivalence between multiplication by a frequency-domain variable and differentiation in the corresponding spatial-domain variable. Since the limits of integration over p do not depend on r, the integral and differentiation operators can be commuted, leaving

$$g(r, t) = \frac{d}{dr} \left\{ \left(\frac{1}{2\pi}\right)^2 \iint_{-\infty}^{\infty} F(p, q) \exp[j(pr + qt)]\, dp\, dq \right\} \tag{5-4}$$

However, the term in the braces is just $f(r, t)$; therefore

$$g(r, t) = \frac{\partial}{\partial r} [f(r, t)] \tag{5-5}$$

Thus the filter jp is just the equivalent of the differential operator in physical space. This example shows how optical filters can be constructed to do mathematical operations. This was an obvious first try; higher-order

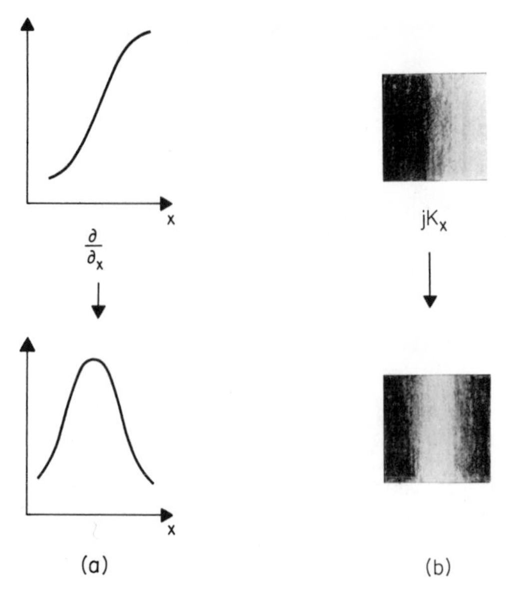

Fig. 5-7. Effect of gradient operations in one dimension. (a) Simple analog effect of gradient. (b) Input mask and output with a jKx filter.

examples are presented in the problems at the end of this chapter. Examples of this type of operation are shown in Fig. 5-7 and Fig. 5-8.

It is quite possible to extend this idea to the question of what equivalent spatial operators can be fabricated through the use of general polynomial-type filter representations. It is obvious that $(jp)^n$ corresponds to an nth order derivative and that $1/(jp)^n$ must then correspond to the inverse operation of integration. It would be most interesting to attempt to find a filter that would correspond to the delaying function $e^{-\beta t}$ in Laplace-transform theory, namely, $1/(s + a)^n$. Fabrication of this more complex filter is difficult and a subject beyond the scope of this book. In general, it is seen that almost any appropriate operation representable by this transform theory can be performed if one can synthesize the appropriate filter.

b. *Deblurring*

One further example of what can be done by this direct approach of spatial filter masks was done by Maréchal [37] when he showed that blurring could be removed by appropriate modification of the low frequencies and appropriate phase reversal. The hypothesis underlying this

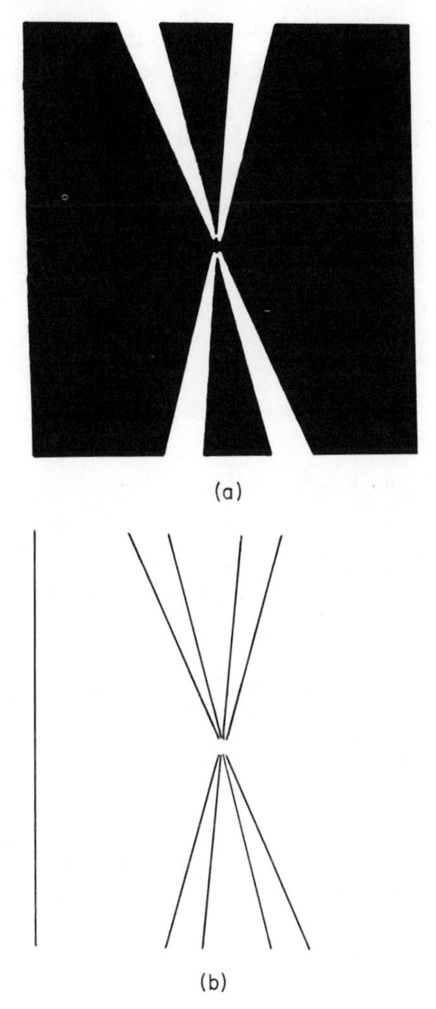

Fig. 5-8. One-dimensional gradient operation on a binary target.

method is that the imaging system that produced the recorded images had certain undesirable features which could be removed or corrected by an appropriately modified viewing system. Basically, blurring occurs in a defocused system leaving a characteristic blur circle. The transmittance function of such a system corresponds to

$$T(\varrho) = 2\,\frac{J_1(\pi A \varrho)}{\pi A \varrho} \tag{5-6}$$

where A is the size of the circular aperture of the input image, and ϱ^2

$= x'^2 + y'^2$ in the viewing plane. A plot of this function is presented in Fig. 2-4. Basically it is desired to suppress the central-region peak and reverse the negative portions. This can be accomplished by using a partially transmitting central disk (size of this disk corresponds to $\varrho_D \simeq 1.22/\pi A$) and a phase reversal plate ($\lambda/2$ retardation plate) over the region of the first negative lobe. Maréchal showed that this procedure resulted in considerable improvement in the corrected image. Others have done this in more sophisticated fashion since the emergence of synthesization procedures that follow the lines of modern communication theory [27].

Complex Filters

Most of these examples have touched on the fact that the filter function $H(p, q)$ is really a complex function. To modify $F(p, q)$ one has to multiply by an appropriate complex function in order to obtain the proper phase and amplitude characteristic in the output plane. The difficulty in obtaining such a complex filter is that, in general, the masks are made from photographic film and the film is exposed in response to the intensity distribution across the plate, where the intensity function at the film plane can be represented by[†]

$$I = UU^* \tag{5-7}$$

This operation of multiplication of the field by its complex conjugate eliminates the phase information.[‡] Thus the mask is real, producing only an amplitude-squared distribution. One must find a way of coding the phase information into the mask so that, by a proper setup, the filter can be made to appear as a complex filter.

Heterodyning

It has long been known that heterodyning enables the extraction of phase from a temporal signal. The analog of this in the spatial domain is interference. Thus it appears plausible that by the mixing of two waves on a

[†] A more complete relationship between fields and film exposure will be developed in the next chapter.

[‡] The reader can demonstrate this to himself by letting $u = Ae^{i\psi}$, where ψ is the generalized phase of the signal and A is the amplitude.

film plate, a quasi-phase recording may be obtained. As in heterodyning, the phase information will be relative phase, or the difference of the phase of the two waves.

This can be demonstrated by letting U in Eq. (5-7) be

$$U = A_1 e^{i\psi_1} + A_2 e^{i\psi_2} \tag{5-8}$$

Substituting in Eq. (5-7) gives

$$\begin{aligned} I = UU^* &= (A_1 e^{i\psi_1} + A_2 e^{i\psi_2})(A_1 e^{-i\psi_1} + A_2 e^{-i\psi_2}) \\ &= A_1^2 + A_2^2 + A_1 A_2 \exp[i(\psi_1 - \psi_2)] + A_1 A_2 \exp[-i(\psi_1 - \psi_2)] \end{aligned} \tag{5-9}$$

This result can be rewritten in a form to demonstrate that I is still real through the use of Euler's relation [14], giving

$$I = A_1^2 + A_2^2 + 2A_1 A_2 \cos(\psi_1 - \psi_2) \tag{5-10}$$

This result is real and can correspond to the amplitude transmission characteristic of a film mask if the appropriate gamma† is used.

To demonstrate that it is possible to obtain a complex signal out of such a filter, use Eq. (5-9), assume a linear relationship between film transmission and intensity profile, and multiply the transmission by the appropriate illuminating signal U_3, where

$$U_3 = A_3 e^{i\psi_3} \tag{5-11}$$

The output signal U_o will be

$$U_o = A_3 e^{i\psi_3} KI \tag{5-12}$$

where K is some appropriate constant. In detail, one has

$$\begin{aligned} U_o = A_1^2 A_3 e^{i\psi_3} + A_2^2 A_3 e^{i\psi_3} + A_1 A_2 A_3 \exp[i(\psi_1 - \psi_2 + \psi_3)] \\ + A_1 A_2 A_3 \exp[-i(\psi_1 - \psi_2 - \psi_3)] \end{aligned} \tag{5-13}$$

If U_1 was the original complex information signal and U_2 was some reference (local oscillator) such as a plane wave, then U_1 can be extracted by adjusting $\psi_2 = \psi_3$ and $A_3 = 1/A_2$. With this specialization, one has

$$U_o = A_1 e^{i\psi_1} + A_1 \exp[-i(\psi_1 - 2\psi_2)] + (A_1^2/A_2 + A_2) e^{i\psi_2} \tag{5-14}$$

† See Chapter VI for definition and more discussion. It basically corresponds to the slope of the *H-D* curve or the exponent in a power-law relationship for transmitted intensities.

where the three terms can be viewed as three complex waves emanating at three different angles. Each angle is determined by the total phase angle associated with each exponent term.† The first wave is the desired wave out of the film. The second is a wave corresponding to the desired complex wave except that it is the complex conjugate of the desired wave and is imaged at an angular position determined by the additional $-2\psi_2$ term in the phase. The last term is a complex term corresponding to the distortions and is in the direction of the illuminating wave ψ_3.

If U_3 is the complex conjugate of U_1, the information wave, note what will happen in Eq. (5-13). Letting

$$U_3 = (1/A_1)e^{-i\varphi_1} \tag{5-15}$$

the output becomes

$$U_o = A_2 e^{-i\varphi_2} + A_2 \exp[i(\psi_2 - 2\psi_1)] + (A_1 + A_2^2/A_1)e^{-i\varphi_1} \tag{5-16}$$

If the coding wave U_2 is a plane wave, then the first term is a plane wave traveling in the reverse direction, the second term is a plane wave at an angle, twice that determined by ψ_1, and the third term is proportional to the information signal where the amplitude has a constant added to it, and the phase is the conjugate phase.

What is important is that for the case that a complex-conjugate signal was present, the output is of a known form and is independent of the form of signal. If U_2 was a spherical wave, then the negative traveling wave would give a focal point on axis and thus one would know the presence of U_1 in the signal simply by a focus on axis. The result has considerable application in the area of character recognition, since the presence of a particular output is indicated in a form independent of the signal and could be the same for a whole set of characters. This process is in general referred to as matched filtering.

Matched Filters

In general, a signal is matched [38] when the impulse response of the filter has the form

$$h(x, y) = ks^*(\xi - x, \eta - y) \tag{5-17}$$

† The concept of angle is easily understood by remembering that the phase term is the dot product of a wave number vector and the position vector ($\hat{k} \cdot \hat{x}$). Thus angle is associated with the projection of the propagation vector into the coordinate system.

where ξ, η, and k are arbitrary constants and k has the added constraint of normalizing the impulse response. Using Eq. (5-17), one can write the output of such a matched filtering system as

$$O(x, y) = \iint S(\xi, \eta)h(\xi - x, \eta - y)\, d\xi\, d\eta$$
$$= k \iint S(\xi, \eta)S^*(\xi - x, \eta - y)\, d\xi\, d\eta = k\, | S(x, y) |^2 \quad (5\text{-}18)$$

where $O(x, y)$ is real and is a maximum signal-to-noise ratio condition. For most cases [13], the matched-filter impulse response $h(x, y)$ is written as the conjugated inverted image $S^*(-x, -y)$.

Using this definition, a method for developing matched filters can be described and used to show how the system is implemented. The purpose of this is to show how an appropriate filter can be used to recognize a certain signal even though it is masked by noise. For example, in this printed page, one might want to find all the places where the letter s is present.

Vander Lugt [39] presented the basic material for this concept, and the following will parallel his work. In discussing any kind of data reduction or detection problem, one needs to establish first some criteria of detection, some basis by which the existence or nonexistence of the signal can be asserted. In the case being considered, the received signal $f(x, y)$ is made up of noise $n(x, y)$ and signal $s(x, y)$. With this received-signal content, the criterion requires that the ratio of peak-signal energy to mean-square noise energy in the output of the processor be a maximum. Since the signal $s(x, y)$ is the object of interest, everything else will be designated as noise $n(x, y)$. This criterion is a classic one, and the solution is well known [39]. To be more concise, assume $n(x, y)$ to be a homogeneous, isotropic random process with spectral density $N(p, q)$, and $s(x, y)$ to be a known Fourier transformable function $S(p, q)$. Now find the optimum linear filter whose impulse response is $h(x, y)$ and whose Fourier transform is $H(p, q)$. For uniform noise power, the solution is

$$H(p, q) = \frac{kS^*(p, q)}{N(p, q)} \quad (5\text{-}19)$$

The constant k is chosen so as to make the magnitude of the $H(p, q)$ less than unity, as required for a passive filter.

Reflecting back on Eq. (5-18), notice that $S^*(p, q)$ can be represented in spatial coordinates by a reversal of direction, giving

$$S^*(p, q) \leftrightarrow s^*(-x, -y) \quad (5\text{-}20)$$

Thus for the case of uniform $N(p, q)$ over all (p, q), one obtains the usual matched-filter case of Eq. (5-18).

With this transfer function, the output of the system can be represented as

$$O(x, y) = (\,) \iint_{-\infty}^{\infty} F(p, q)H(p, q) \exp[i(px + dy)]\, dp\, dq \qquad (5\text{-}21)$$

where Fig. 5-9 represents the system.

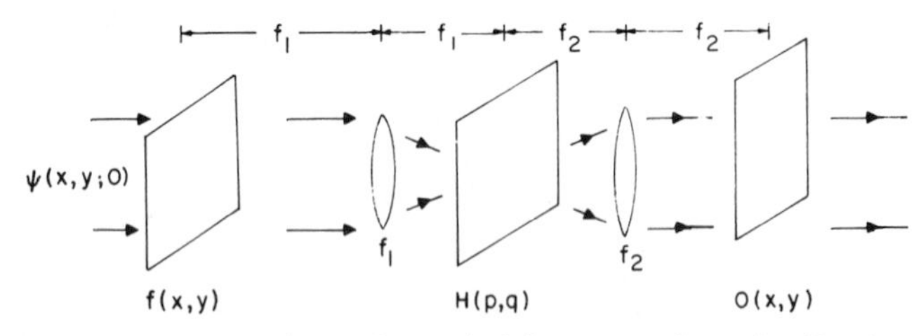

Fig. 5-9. Simplified system layout for optical data processing using Fourier-transform filtering. The input mask distribution is $f(x, y)$; the Fourier plane mask distribution is $H(p, q)$; and the output amplitude distribution is $O(x, y)$.

The problem now reduces to finding the $H(p, q)$ which is a ratio of two spectra. In Chapter VI the method of how to form ratios of spectra on photographic film will be given. Suffice it to say that one can cascade two filters together to form the desired filter. Since uniform noise has been assumed, the $1/N(p, q)$ is just a constant normalizing the conjugated signal spectrum.

The problem now reduces to finding $S^*(p, q)$, which, because of the conjugate, means that one cannot simply record the output of a Fourier transform of the signal. Recording, in this simple case, would give a nonzero real filter.

As mentioned before, some indirect procedure to record the phase, such as an interferometer, must be used. Vander Lugt has shown that a Mach–Zehnder interferometer will work, since it enables the mixing of two signals from a single source. A typical Mach–Zehnder is shown in Fig. 5-10. Now the recorded output of one Mach–Zehnder system is

$$\begin{aligned} G(p, q) &= | R(p, q) + S(p, q) |^2 \\ &= | R(p, q) |^2 + | S(p, q) |^2 + R^*(p, q)S(p, q) + R(p, q)S^*(p, q) \end{aligned} \qquad (5\text{-}22a)$$

or

$$\begin{aligned} G(p, q) = {}& | R(p, q) |^2 + | S(p, q) |^2 \\ &+ 2 | R(p, q) | \, | S(p, q) | \cos[\phi(p, q) - \theta(p, q)] \end{aligned} \qquad (5\text{-}22b)$$

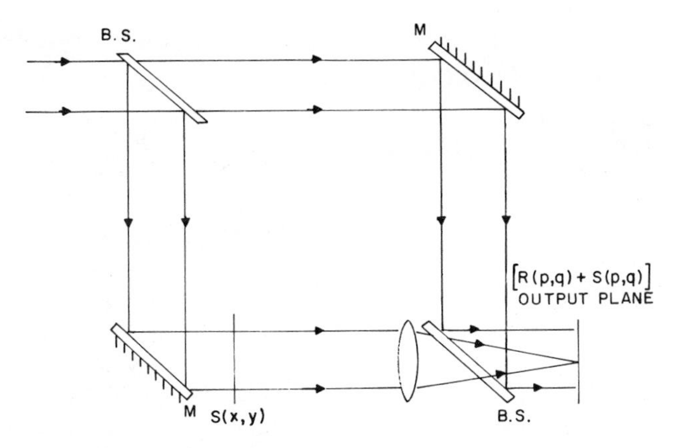

Fig. 5-10. A Mach–Zehnder interferometer setup for matched-filter generation. The input is $S(x, y)$, and the output is $R(p, q)$ plus $S(p, q)$.

where

$$R(p, q) = |R(p, q)| \exp^{j\phi(p,q)} \tag{5-22c}$$

and

$$S(p, q) = |S(p, q)| \exp^{j\theta(p,q)} \tag{5-22d}$$

This $G(p, q)$ is a nonnegative function that allows determination of $\theta(p, q)$ by proper adjustment of $\phi(p, q)$.

The transmission through a composite filter using $1/N(p, q)$ as the first in the composite set gives

$$\frac{G(p, q)}{N(p, q)} = A(p, q) + H^*(p, q) \exp[-j(bp + cq)] + H(p, q) \exp[j(bp + cq)] \tag{5-23}$$

where

$$A(p, q) = \frac{|H(p, q)|^2 + |S(p, q)|^2}{N(p, q)} \tag{5-24a}$$

$$H(p, q) = \frac{kS^*(p, q)}{N(p, q)} \tag{5-24b}$$

$$\phi(p, q) = bp + cq \tag{5-24c}$$

and b, c are constants. The third term in the equation is the desired filter function. However, it is coupled to two other terms. The problem is to separate it from these other terms.

If this filter is inserted back in the optical processor at the focal of $f(x, y)$, then the output can be written as

$$O(x, y) = (\,) \iint_{-\infty}^{\infty} F(p, q)A(p, q) \exp[j(px + qy)]\, dp\, dq$$
$$+ (\,) \iint_{-\infty}^{\infty} F(p, q)H^*(p, q) \exp\{j[(x - b)p + (y - c)q]\}\, dp\, dq$$
$$+ (\,) \iint_{-\infty}^{\infty} F(p, q)H(p, q) \exp\{j[(x + b)p + (y + c)q]\}\, dp\, dq \tag{5-25}$$

If the terms of Eq. (5-25) are examined, it is seen that the first term is nothing more than distortion and appears on axis. The second term appears off axis at $x = +b$, $y = +c$. The third term located at $y = -b$, $x = -c$ is the desired output from the optimal filter.

Since the Mach–Zehnder interferometer is a single-plane device and since the three terms can be separated in a single plane, one can set c equal to zero. The only remaining problem is that b be large enough to prevent information overlap with the on-axis distortion terms. Thus $b \geq B$ is a condition related to the length and content of the signal in x.

The terms in Eq. (5-25) can be written in the convolution format because the inverse transform of each product of two transforms is a convolution. More specifically, if $N(p, q)$ is a constant,

$$r(x + b, y) = (\,) \iint_{-\infty}^{\infty} f(u, v)s^*(x + b + u, y + v)\, du\, dv \tag{5-26}$$

which is a cross-correlation integral, and

$$r(x - b, y) = (\,) \iint_{-\infty}^{\infty} f(u, v)s(x - b - u, y - v)\, du\, dv \tag{5-27}$$

which is a convolution integral. Thus the desired filtering process can be viewed as a cross-correlation process, and it appears off axis as shown in Fig. 5-11.

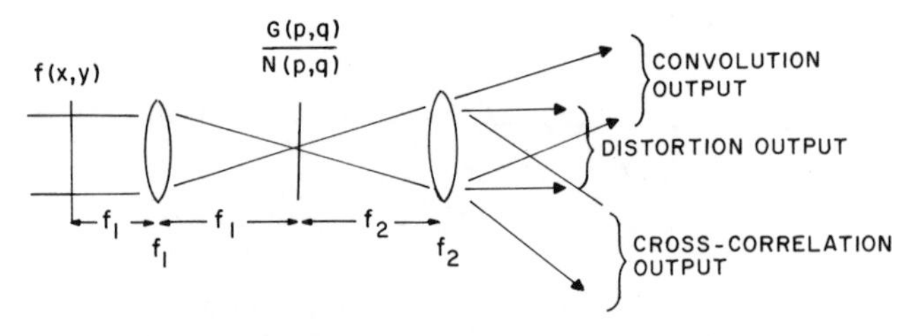

Fig. 5-11. System outputs for matched filters.

Weiner–Kolmogorov Estimation Filter

Another alternative for optimally detecting signal from noise follows the establishing of another detection criteria. In this case, the criteria for optimization will be that the mean-square error be a minimum [18]. The error that is being referred to in this case is the difference between the estimated signal $s'(x, y)$ and the actual signal $s(x, y)$.

This approach then specifies a filter, which, when applied in the processor, gives a linear estimate $s'(x, y)$ of the signal present in the input signal $f(x, y)$, where the input signal is the sum of signal $s(x, y)$ and the noise $n(x, y)$. The linear estimate can be represented mathematically as

$$s'(x, y) = \iint_{-\infty}^{\infty} f(x - u, y - v)h(u, v)\, du\, dv \tag{5-28}$$

where $h(u, v)$ is the impulse response of the desired system. It is the objective then to find $h(u, v)$. The criterion is to minimize the mean-square error ε, where

$$\varepsilon(x, y) = s'(x, y) - s(x, y) \tag{5-29}$$

The mean-square error (MSE) is defined as the expectation (E) or average of the squared error,

$$\text{MSE} = E[| s' - s |^2] \tag{5-30}$$

It has been shown [21] that the minimum of Eq. (5-30) is obtained when the error is orthogonal to the received signal in an average sense,

$$E[(s' - s)f] = 0 \tag{5-31}$$

That is, the average of the inner product of the error and the input signal is zero. Expanding this statement and averaging gives

$$E[s'f - sf] = E[s'f] - E[sf] = 0 \tag{5-32}$$

or rewriting, using the linear estimate expression, leaves

$$\begin{aligned} E[sf] &= E\left[f \iint_{-\infty}^{\infty} f(x - u, y - v)h(u, v)\, du\, dv\right] \\ &= \iint_{-\infty}^{\infty} E[f(x, y)f(x - u, y - v)]h(u, v)\, du\, dv \end{aligned} \tag{5-33}$$

However, the term inside the integral is just the autocorrelation of f with itself and the left-hand side is just the cross-correlation of s with f, R_{sf}, giving

$$R_{sf}(x, y) = \iint_{-\infty}^{\infty} R_{ff}(x - u, y - v)h(u, v)\, du\, dv \tag{5-34}$$

which, if expanded, yields

$$R_{ss}(x, y) + R_{sn}(x, y) = \iint_{-\infty}^{\infty} [R_{ss}(\,) + R_{nn}(\,) + 2R_{sn}(\,)]h(u, v)\, du\, dv \tag{5-35}$$

If the signal and noise are assumed to be uncorrelated ($R_{sn} = 0$), then these equations reduce to

$$R_{ss}(x, y) = \iint_{-\infty}^{\infty} [R_{ss}(x - u, y - v) + R_{nn}(x - u, y - v)]h(u, v)\, du\, dv \tag{5-36}$$

This result is written in the functional form of a convolution; therefore, one can take the Fourier transform of both sides and get a solution for the filter transfer function,

$$H(p, q) = \frac{S_{ss}(p, q)}{S_{ss}(p, q) + S_{nn}(p, q)} \tag{5-37}$$

The terms on the right-hand side are the power spectral density $S_{ss}(p, q)$ of signal (intensity distribution of the Fourier spectrum of the signal) and the power spectral density of the total received signal $f(x, y)$.

This result is important, because it has the added advantage that it is a real filter that is realizable on film without any special modulation or interferometry techniques. It suffers, however, in that it does not totally separate noise from the signal; it only suppresses the noise relative to the signal present.

This technique has been used to count reticulated blood cells in the presence of noise [40] even when the population of reticulocytes has been as low as 1000 reticulocytes per 1,000,000 red blood cells present (erythrocytes and reticulocytes).

An Application of Weiner–Kolmogorov Filtering

An illustration of the Weiner–Kolmogorov (W–K) estimation technique is shown in Figs. 5-12 through 5-20. This application concerned itself with the recognition and counting of immature red blood cells (reticulocytes) [40]. Figure 5-12 shows examples of the kind of cells that were being sought. The slightly larger cells in the center of the field with the dark material

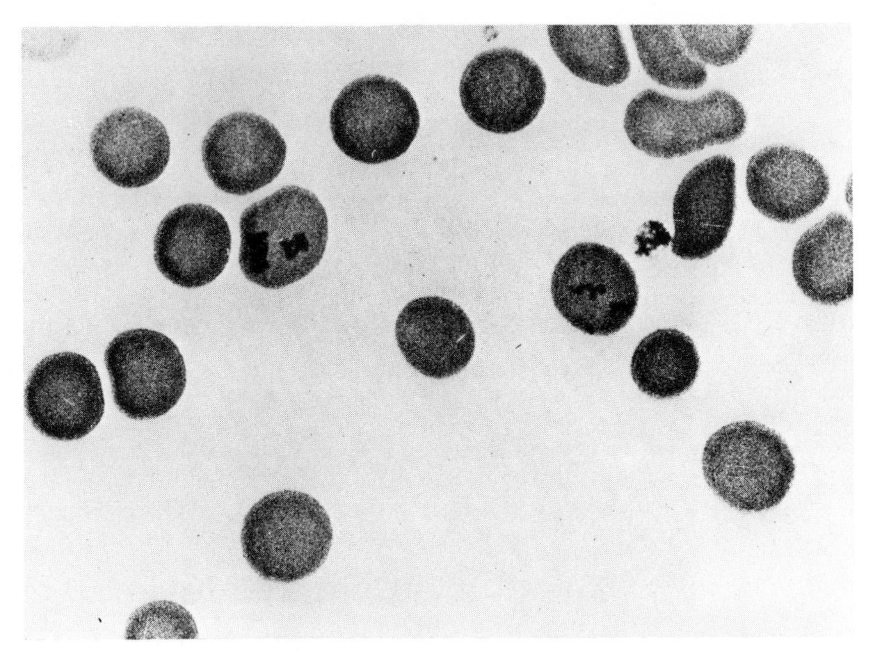

Fig. 5-12. A section from a blood slide showing reticulocytes and erythrocytes along with system stain artifact.

(reticulum) inside are the reticulocytes. The other cells are the mature cells (erythrocytes) and in this application are the major constituent of the noise in the process.

A diagram of the optical system is presented in Fig. 5-13. A laser, with the appropriate collimating optics, illuminates a standard blood slide† which in turn illuminates the W–K filter that has been placed in the focal plane of the lens L_3. Samples of the unfiltered focal-plane patterns [Fourier transforms of $f(x, y)$] are shown in Fig. 5-14. The last lens, L_4, performs

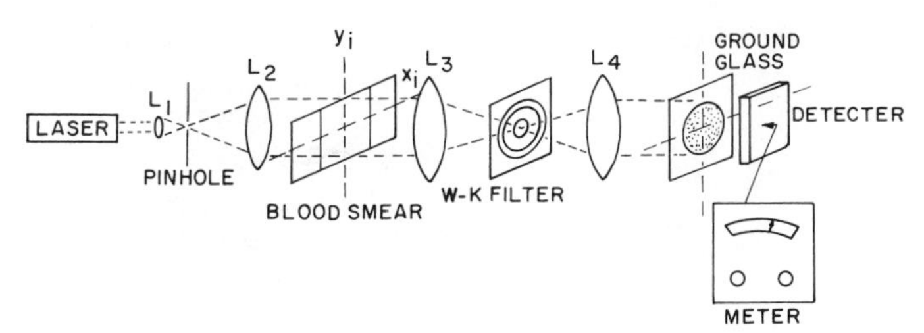

Fig. 5-13. A system layout for an optical Weiner–Kolmogorov estimation counter.

† See reference [40] for details on the slide preparation.

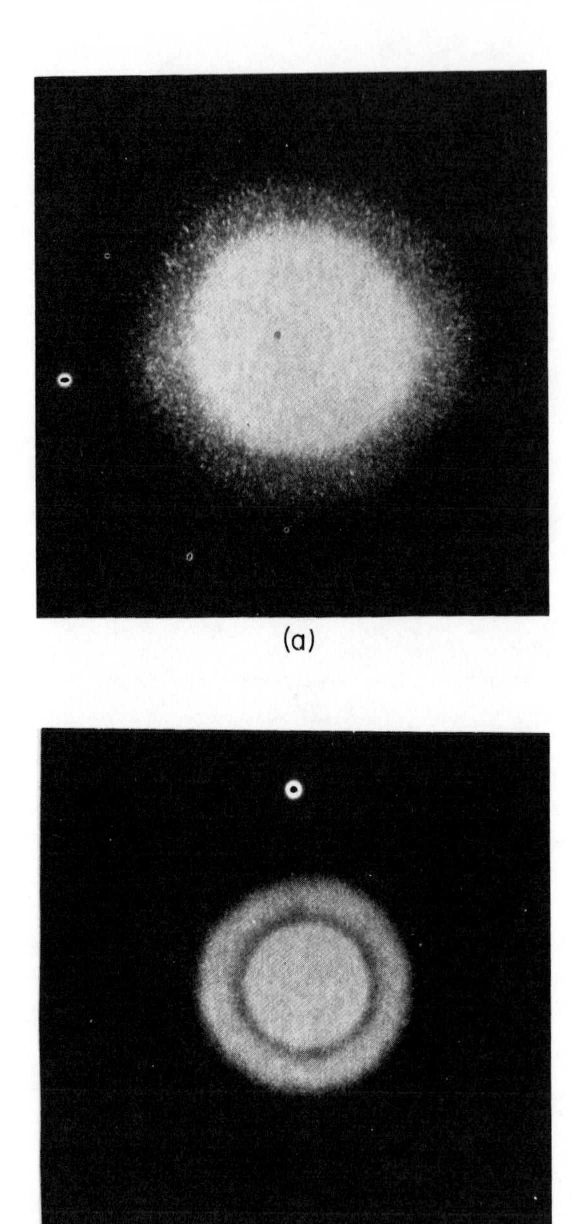

Fig. 5-14. (a) Focal-plane distribution for a 60-percent reticulocyte count. (b) Focal-plane distribution for a normal 1- to 2-percent reticulocyte count.

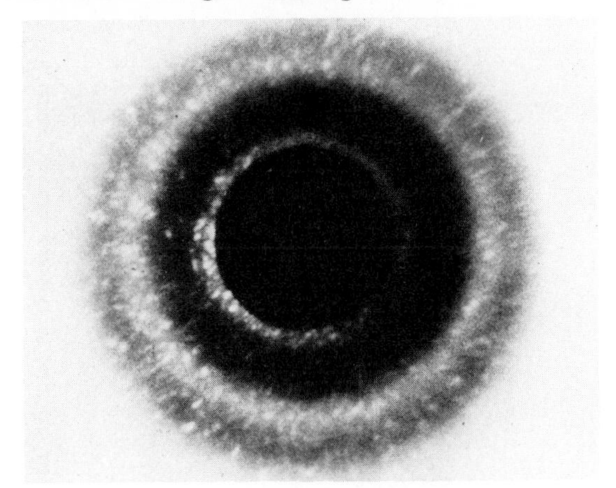

Fig. 5-15. A typical Weiner–Kolmogorov filter for reticulocytes.

the inverse transform, and the detector responds to the total output of the system integrated by lens L_4.

A sample of the W–K filter used for reticulocytes is shown in Fig. 5-15. The center region is purposely blocked to enhance the higher spatial-frequency weighting. Figure 5-16 shows the meter outputs for a random

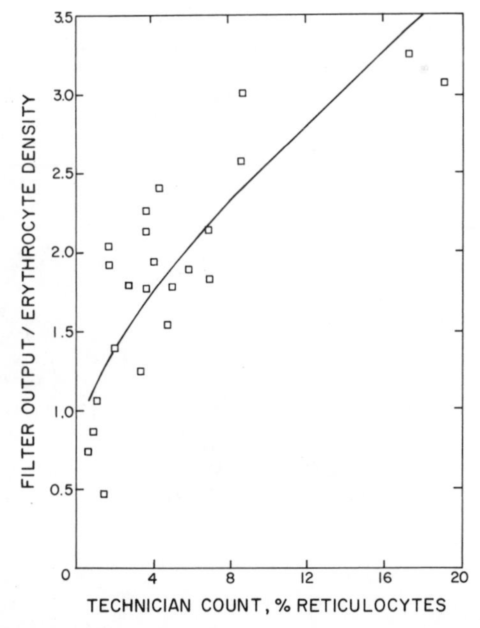

Fig. 5-16. Normalized system output versus technician counting for reticulocytes with calibration curve superposed.

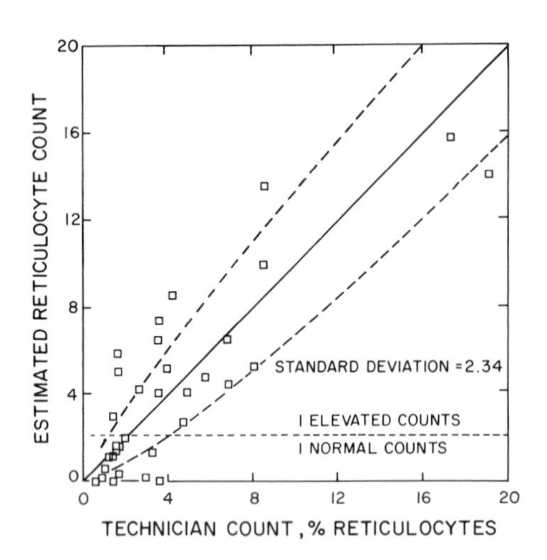

Fig. 5-17. Estimated count versus technician count with standard deviation data superposed.

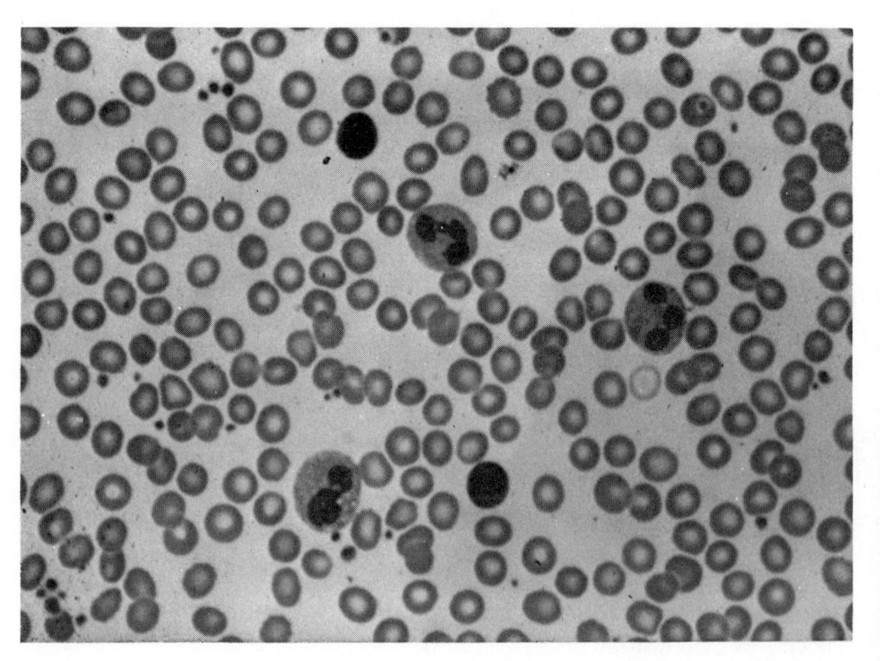

Fig. 5-18. A blood slide section showing white cells—monocytes and lymphocytes.

sample of blood slides. The solid line represents a least-squares fit to the data. This curve then constitutes the calibration curve for a second random set of blood slides which are counted for reticulocytes. These data are presented in Fig. 5-17.

This procedure has been further extended to white cells as shown in Fig. 5-18. The cells in the center field are white cells, and it is desired to

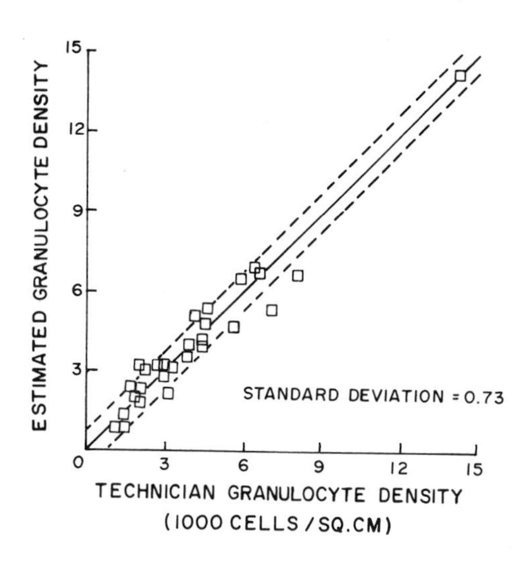

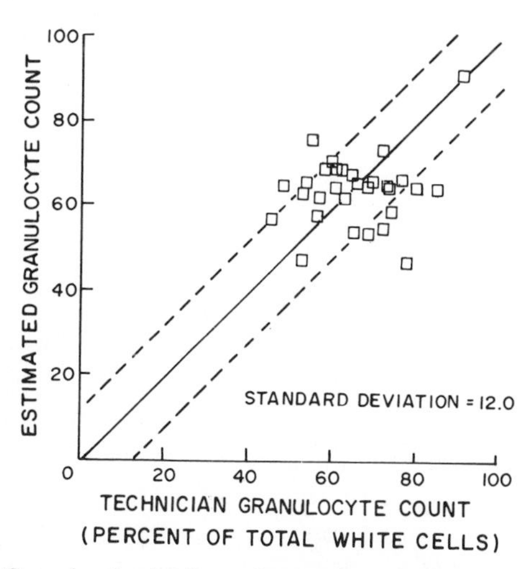

Fig. 5-19. (a) Granulocyte density estimates for a Weiner filter. (b) Granulocyte differential counts.

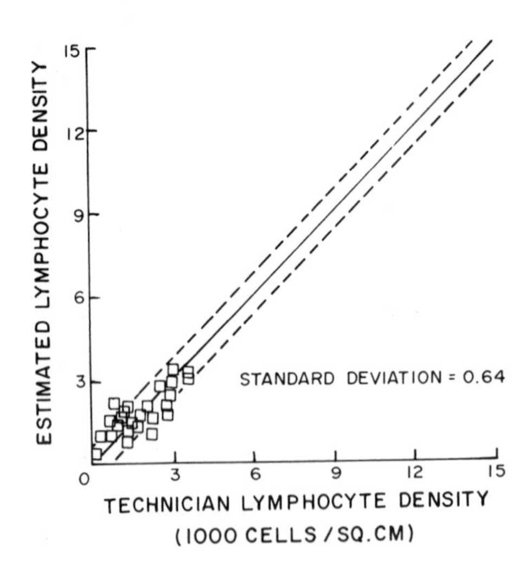

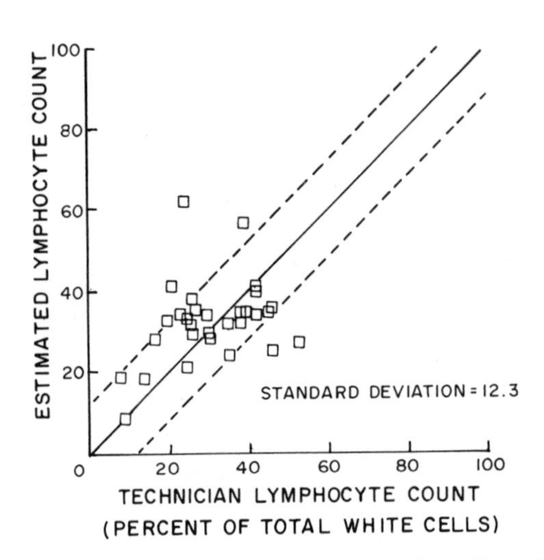

Fig. 5-20. (a) Lymphocyte density estimates for a Weiner filter. (b) Lymphocyte differential counts.

recognize the type, density, and percentage present in a given sample. This technique has been specifically applied to lymphocytes and granulocytes. The results of these samples are shown in Fig. 5-19 and Fig. 5-20.

Presentation of these brief examples is meant only to be illustrative of the power of this technique. There are many other applications that can be thought of, and the general application has barely been touched.

Synthetic Aperture Radar

a. *Concept*

One of the earliest successful applications of coherent optical data processing was in the area of synthetic aperture radar [8, 25, 38–41]. This is a system in which an equivalent large aperture is obtained by storing data on film at many sample points and then resynthesizing the image in an optical system that corrects for inherent aberration and tilt.

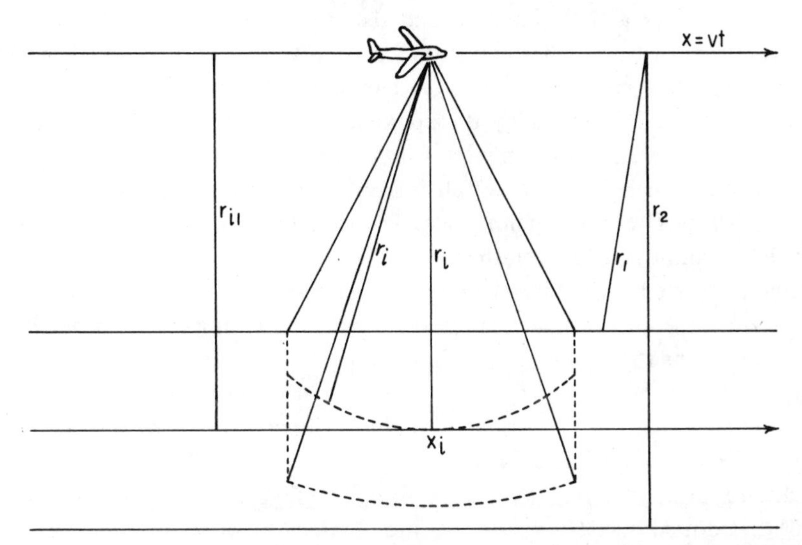

Fig. 5-21. An airborne synthetic aperture radar system collecting data for later processing.

Figure 5-21 illustrates the basic setup for data collection. Basically this system is chosen because a more conventional radar imaging system has an inherent limit on the resolution which is on the order of $\lambda_r R/D$, where R is the range, D is the antenna linear dimension, and λ_r is the radar wavelength. Thus a narrow-beam system is basically resolution limited. Another way of thinking of this narrow-beam concept is that to achieve the needed resolution, the antenna system must have a large dimension D made up of many elements that are appropriately phased together. The size of such an array would be so large that it could not be transported in a practical aircraft. Thus think of this approach as having the processing in the antenna system.

An alternative to this narrow-beam system is to accept that such an array cannot be built with inherent processing capability, but plan to do the processing from many elements after the fact. That is, in essence, where the term "synthetic aperture" comes from. What is done is that the return from a coherent radar pulse is sampled at many points along a given flight path, and then the information from each of these sample points is stored on film for later processing.

In Fig. 5-21 the basic system is shown. A coherent pulse having a wide-beam characteristic is sent out over a region, and a synchronized receiver is turned on to observe the return from the illuminated region. The receiver is turned on for a short interval, and the receiver output is used to drive a CRT vertical trace. The intensity of the CRT trace is modulated by the signal, and this information is stored on film. The resolution of such a system is basically limited by the apparent size of the illuminated region, or $2R\Omega$, where Ω is the angular beam width. The minimum resolvable dimension is then $\lambda R/2\Omega R$, which for diffraction-limited antenna systems is $\frac{1}{2}D$, where D is the antenna size. Thus for centimeter wavelengths the resolution limit is a few centimeters.

Since the receiver is turned on only after a delay and is turned off in a short time, the area sampled can be equated to observing the region defined by r_1 and r_2 and the illuminating dimension ΩR. During the instant of time that a return is being recorded for a radius r_i, it is important to realize that all points with the same apparent radius will contribute to the received signal. Thus at any time t_i in the sampling interval, the displayed intensity is due to all points lying on a segment of an ellipse. This is shown in Fig. 5-21 as a dotted line. The range data are obtained by pulse timing.

If the plane of the earth were perpendicular to the propagation path, the locus of equivalent radii would be arcs of a circle. An awareness of these curved paths is important in visualizing what kind of corrective optics is necessary in the subsequent processor.

b. *Data Storage*

The data received during one sample interval are stored on film that is transported past the CRT trace. A representation of the film storage system is presented in Fig. 5-22. Each vertical trace corresponds to one time sample. The vertical direction corresponds to range; the closest point r_1 appears at the bottom; and the maximum range r_2 is at the top.

Also shown in this figure is the track of a point target. As the aircraft flies by, a given target will be first recorded or indicated by a radius that is a maximum; that is, it just barely becomes visible. As the aircraft continues to advance, the range to target becomes less until the minimum perpendicular range r_i is obtained. Then the aircraft passes the closest point of approach, and the range increases again. This curved track is really a segment of a hyperbola which will later be approximated by a parabola. The boundary curves or asymptotes of the hyperbolic family are straight lines representing targets on the path of the aircraft. The track would in fact not appear as a series of dots, but each dot would be made up of a main lobe and visible side lobes depending on the reflectance characteristics of the target. The arcs in this description are exaggerated in curvature to aid the description of the process.

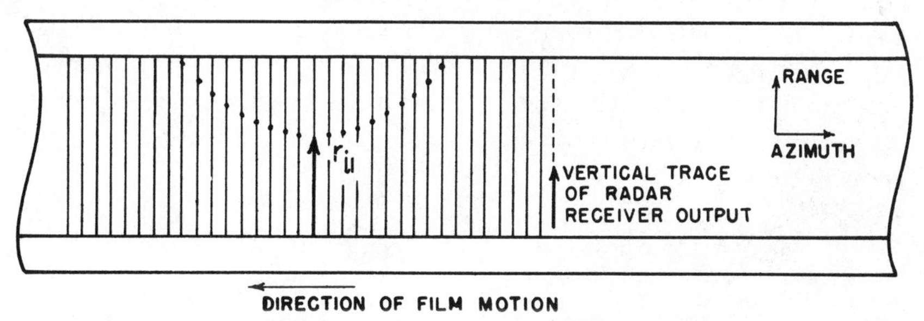

Fig. 5-22. Recording format for radar images on film.

It is at this point that one can intuitively identify the correcting optics needed for the system processor. First, note that a far-field spread representation of the target is obtained when the aircraft passes by it; thus in one dimension† the target is represented by a spectrum or Fourier transform. In the other dimension a simple range condition is obtained. Thus the form of a one-dimensional transforming lens (cylindrical) is suggested.

This picture is complicated, however, by the fact that the spectral path is curved. This means that at various ranges, the radius of curvature must change. Hence a simple cylindrical lens is not adequate. The curved nature of the path could be corrected by having a wedge whose angle was a function of azimuth offset. That is, as the target first enters the illuminated region, its position can be referenced to the closest point of approach angle by an azimuth angle. It is at this position that the maximum amount of bending is required or the maximum wedge angle is needed.

† The dimension referred to here is along the hyperbolic track.

Thus if a variable angle wedge and a cylindrical lens were properly combined, the target can be reimaged. It just so happens that this configuration is known as a conical lens, and it is characterized by a V-shaped structure where the radius of curvature increases as one moves from the vertex of the *V*.

Physically then, it is easy to see why the large amount of resolution is obtained. The target is described by many sample points, each having the appropriate phase weighting to give a complete one-dimensional spectral characteristic.

c. *Mathematical Representation*

Following the format of Cutrona *et al.* [42], a mathematical model of the data format can be derived. A single-point source target will be assumed located at a point (x_i, r_i). More complicated targets can be represented by an appropriate sum over target representations, much as is conceptually done in Huygens' representation of a radiating aperture. The received signal $S_i(t)$ can be written as

$$S_i(t) = \alpha_i \exp[j\omega_r(t - 2r/C)] \tag{5-38}$$

where ω_r is the radar frequency in radians per second, α_i represents the complex amplitude containing antenna and target pattern characteristics. If x is the position of the aircraft ($x = 0$ at $t = 0$), x_i is the position of the target along a parallel path, and $r_i \gg (x - x_i)$, then r can be represented through the binomial expansion as

$$r = [r_{i1}^2 + (x - x_i)^2]^{1/2} \cong r_{i1} + \frac{(x - x_i)^2}{2r_{i1}} \tag{5-39}$$

Substituting this in Eq. (5-38) and using the fact that time t and position x are related through the aircraft velocity v_a by $t = x/v_a$, $S_i(t)$ can be written as

$$S_i(t) = \alpha_i(x_i, r_i) \exp\left[j\left(\frac{\omega_r x}{v_a} - \frac{4\pi r_{i1}}{\lambda_r} - \frac{2(x - x_i)^2}{\lambda_r r_{i1}}\right)\right] \tag{5-40}$$

This radar signal is synchronously detected or down-shifted in frequency such that

$$S_i(t) = |\alpha_i(x_i, r_i)| \cos\left[\frac{\omega_c x}{v_a} - \frac{4\pi r_{i1}}{\lambda_r} - \frac{2}{\lambda_r r_{i1}}(x - x_i)^2 + \phi_i\right] \tag{5-41}$$

This real signal intensity modulates a CRT trace which yields ultimately a film transparency that has, as a transmission function, $t(x, y)$,

$$t(x, y) = t_b + \xi \mid \alpha_i(x_i, r_i) \mid \cos\left(\frac{\omega_c x}{v_f} - \frac{4\pi r_{i1}}{\lambda_r} - \frac{2\pi}{\lambda_r r_{i1}}\left(\frac{v_a x}{v_f} - x_i\right)^2 + \phi_i\right) \tag{5-42}$$

where t_b is a film bias to allow for negative portions of the cosine function, x is now the position on the film transport, and v_a/v_f is the scaling factor for the change in path position representation.[†] It is important to remember that this is the representation for a single-point target. Several targets would entail the use of a sum over the i index.

This film transparency $t(x, y)$ can be viewed as a complex grating. It will scatter waves if illuminated by a coherent optical wave in both the forward and reverse directions. The scattering effect can be seen by decomposing the cosine function into two exponentials using the Euler relations. This is similar to what was done earlier in this chapter. First, however, this expression can be simplified by moving constant phase terms like ϕ_i and $4\pi r_{i1}/\lambda_r$ into a new amplitude function.

The term of interest is the real-image scattering term which is represented by

$$t_r(x_i, y_i) = \frac{\eta}{2} \alpha_i'(x_i, y_i) \exp\left[j \frac{\omega_c x}{v_f}\right] \exp\left[-j \frac{2\pi}{\lambda_r r_{i1}} \left(\frac{v_a}{v_f}\right)^2 \left(x - \frac{v_f}{v_a} x_i\right)^2\right] \tag{5-43}$$

where the terms of the exponent function have been intentionally separated by making a product of exponents. The scattering terms are shown in Fig. 5-23.

Physically, the two terms correspond to the wedge and cylindrical lens identified earlier. The linear phase change inherent in the first term corresponds to a wedge. The second term can be related to the equation for a positive cylindrical lens at position $x = x_0$ by

$$t_{cyl}(x_i, y_i) = \exp\left[-j \frac{k_0}{2f_i} (x - x_0)^2\right] \tag{5-44}$$

[†] The velocity v_f is the velocity of the film transport.

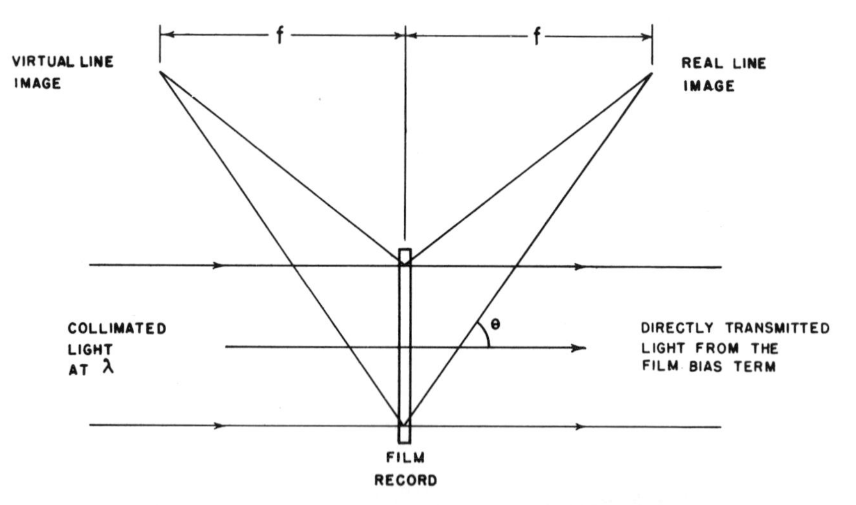

Fig. 5-23. Twin image formation from the film record.

If the focal length F_i is set equal to

$$f_i = \frac{1}{2} \frac{\lambda_r}{\lambda_0} \left(\frac{v_\mathrm{f}}{v_\mathrm{a}}\right)^2 r_{i1} \tag{5-45}$$

the location of this particular cylindrical lens is given by

$$x = (v_\mathrm{f}/v_\mathrm{a})x_i \tag{5-46}$$

which correlates with the intuitive motion that the radius of curvature of the lens had to change as a function of the vertical position of the trace.

d. *Coherent Imaging System*

A system which would do such an operation is shown in Fig. 5-24. The conical lens is shown adjacent to the film record. The cylindrical lens forms a focal-plane image in the range direction to match the focal-plane image formed by the conical lens in the azimuth direction. The spherical lens then reimages this focal-plane image into a real-image plane that displays a very high-resolution image of a radar map. In the Cutrona paper [42] and in other references [28, 41], sample maps are provided. The resolution is excellent and far exceeds any direct imaging radar systems.

The importance of encoding the real image on a spatial sideband was emphasized in the discussion of Vander Lugt matched filters [39]. The importance is evidenced here again, as without the spatial modulation provided by the interferometric record, one would not have been able to separate the various images, and in particular the record would be dominated by the on-axis noise terms.

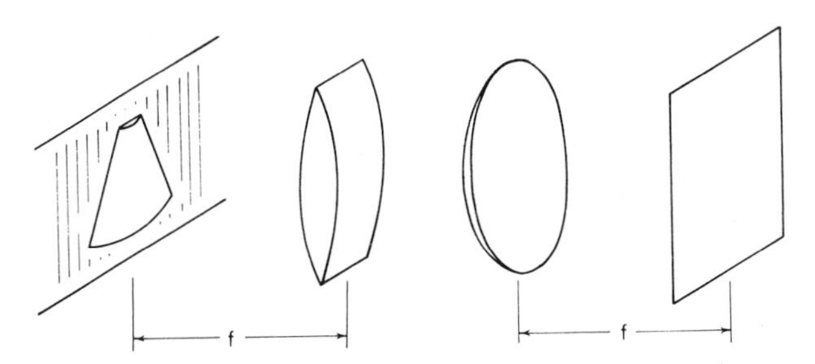

Fig. 5-24. An optical system for imaging the stored aperture data using a special conical lens.

Problems

1. Using arguments similar to those for the xy high-pass filter, synthesize a filter that would be y low pass, x high pass.

2. What does the filter $H(p, q) = -p^2$ correspond to in physical space?

3. What does the filter $H(p, q) = 1/jp$ correspond to in physical space?

4. Calculate the size of the first Airy disk for a simple Fourier system if the focal length of the lens is 170 mm and the blood cell is on the average 10 μ in diameter. (Assume $\lambda = 0.5$ μ.)

5. Starting with Eq. (5-36) show the details leading to Eq. (5-37).

6. Using the Fourier transform of a square grating in a focal plane, describe with band-pass filters how one would obtain an output image whose amplitude distribution was like a sine grating.

7. If one observes the letters X, F, K, V, M, T, Y, L, H, W, they all have certain things in common; i.e., they are all made out of straight line segments at specific angles. The Fourier transforms of certain members of the set will be very similar and in fact very difficult to distinguish.

(a) Separate the letters into subsets having similar spectrums.

(b) Sketch each subset transform pattern.

(c) Design binary filters that will enable one to distinguish one subset from another.

8. If the complex number j can be represented by $e^{j(\pi/2)}$ and $-j$ by $e^{-j(\pi/2)}$, design a filter for jp using a filter $|p|$ and an appropriately mounted half-wave plate.

9. The letters X and V are very similar in an image sense. Their focal-plane images have a subtle difference. What is it? What filter could convert an X to a V or vice versa?

Chapter VI

INTERFACE DEVICES

Photographic Recording

Photographic films and plates are important in optics, since they serve as input devices, output devices, and filter masks in optical systems. As input devices, they are used as masks for the input-image information. As output devices, they are used primarily to record an output image, but can also be used as limiting or time-dependent filters in some processes. As filter masks, they are used to modify signals transmitted by a given system plane.

In referring to photographic films and plates, the distinction is primarily concerned with the backing material used for holding the emulsion. Photographic film employs a plastic backing, most commonly acetate, and photographic plates use glass. Generally, photographic plates have a finer-grain emulsion and are used for spectroscopic work.

The emulsion is generally a gelatin material in which many tiny silver halide grains are suspended. Under exposure to light, the tiny silver halide sites absorb optical energy and undergo a chemical change. In this change, the silver ions disassociate and form little development centers of metallic silver. Under a subsequent process, these development centers cause the whole grain to become metallic silver. Grains without centers do not form into metallic silver, but remain as unexposed sites. The fixing process, which follows development, removes these undeveloped grains and leaves the sites of metallic silver. Since these metallic silver sites do not transmit visible light, the exposed region becomes opaque. This opacity is a function of the number of silver grains, and hence opacity is a function of the density of silver deposited in the developed sites.

To calculate the density D, a convenient representation needs to be established. Density is not the parameter used directly in the system modeling. Transmission T is a much more appropriate parameter, since it relates directly with the function the film is required to perform. In addition, since transmission can vary over such a large dynamic range while D varies a small amount, a standard technique of logarithms is utilized. Thus, the relationship between D and T is defined as [45, 46]

$$D = \log(1/T) \tag{6-1}$$

An inverse relationship on T is used because when the density D is large, the transmission is small. The transmission T is the ratio of intensities. This ratio is the intensity of light that an object passes divided by the intensity of the light incident on it.

The description and measurement of the photographic properties of emulsions are the science of sensitometry. In sensitometry, much of the data are empirical and can best be discussed in terms of characteristic curves as shown in Fig. 6-1. These curves are referred to as Hurter–Driffield curves [45].

To discuss this curve, it is necessary to define several terms. First is the idea of exposure. Since the silver ion disassociation from the halide ion is brought about by a deposit of optical energy (W-sec), the input energy density (W-sec/cm²) must somehow be related to the input intensity

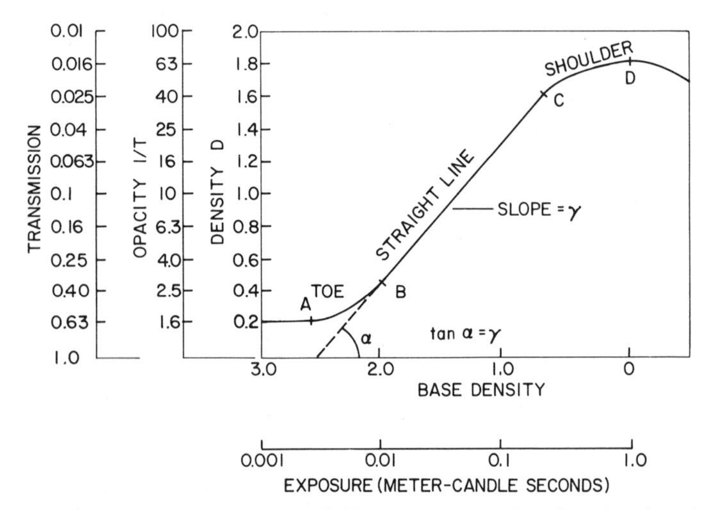

Fig. 6-1. Characteristic Hurter–Driffield (H–D) curve for film showing deposited density as a function of log exposure.

(W/cm²). The term associated with this input energy density is exposure. It can be extrapolated from the reciprocity law of Bunsen and Roscos [45], which states that the product of photochemical reactions is dependent on the total optical energy used. Thus, exposure E is defined as

$$E = \int_0^T I_i \, dt \qquad (6\text{-}2)$$

where I_i is the illuminating intensity, T is the total exposure time (sec), and t is time in seconds. Most often, the intensity is constant over the exposure level so Eq. (6-2) reduces to

$$E = I_{i0} T \qquad (6\text{-}3)$$

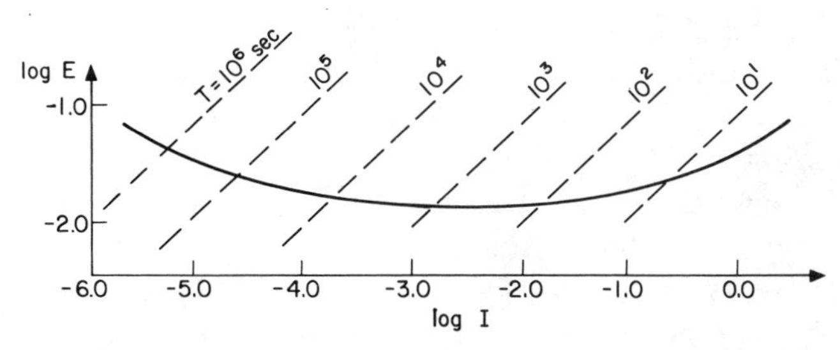

Fig. 6-2. Characteristic curve showing reciprocity failure. The constant parameter is deposited density.

Most often, the two factors in the exposure equation are independent. The cases where this is not true are for very high and very low intensities. In these intensity extrema, photographic materials show a loss of sensitivity. The decrease in sensitivity is known as the "reciprocity effect," or more precisely the "failure of the reciprocity law." The effect is shown in Fig. 6-2, where the curve is plotted for a constant deposited density D. The reciprocity law, if ideal, would require the same amount of radiant energy to produce the same density, at constant exposure, independent of the intensity level. For many materials, however, logarithm of the exposure log E increases for the same density at high and low intensities. It is a slowly varying characteristic, but sufficient to make log E change by a factor of two when the logarithm of the intensity, log I, changes by six decades. The minimum is a region of maximum sensitivity.

The characteristic loss of sensitivity is particularly important at large intensities since the advent of lasers, where intensities can be extremely high (MW/cm²) and exposure times very short (psec). The loss of sensitivity at low intensities is important to consider in astronomy. In astronomical recordings intensities are very small ($\mu W/cm^2$), thus requiring very long exposure times (ksec). More detail is provided in Fig. 6-2. The limitations described above make considerations of reciprocity failure important in film selection. It should be noted that the effect is temperature dependent, and some improvement in the sensitivity can be achieved by cooling the emulsion. The effect, however, shows very little spectral dependence when the same density and exposure times are used.

Exposure was defined as the product of intensity and time. It follows then to question the effect time continuity plays in a given exposure. Given a system operating at constant illuminating intensity, the first question has to do with whether the exposure time is continuous or segmented. It has been found experimentally that intermittent exposure times produce different effects until a critical interval spacing is reached. Above this segmentation, the intermittent exposure is about the same as the continuous exposure. This is known as the "intermittency effect" and the critical interval is about $T_e/100$.[†]

The effects of varying the intensity can also be examined. If the intensity in a given exposure is broken into parts of different magnitude, it is found that a brief, very intense pulse of light, followed by a longer, more moderate pulse of light, do not add in a simple fashion. The first pulse desensitizes the emulsion effectively toward the second pulse. If the two pulses of light cover some of the same spatial points on the film, then those portions with both intensities present will appear reversed with respect to the first pulse upon development, yielding a positive. This effect is known as the "Clayden effect." This phenomenon is frequently observed in photographs of lightning flashes. Thus the effect has derived the name, "black lightning."

The effect known as "solarization" is the decrease in density for exposures that go beyond the shoulder of the H–D curve. This is equivalent to a reversal of the image. This effect is shown in Fig. 6-1. Basically sites are caused to reassociate with the halides.

The "Herschel effect" is the ability to erase parts of an exposure by exposing the emulsion to longer-wavelength radiation. To counter this effect, emulsions are often desensitized to the red or infrared by appropriate dyes. The energy required at longer wavelengths to accomplish this erasure is several orders of magnitude greater than the initial energy required to form the original latent image.

[†] T_e is the total exposure time. Given that $IT_e = I \sum_i^N (T_e/N)$, the intermittency effect disappears for $N > 100$.

Resolution

Resolution in photographic emulsions is a concept which is not clearly defined. Not only is the term itself subject to too many definitions, but there are many other terms that have related meanings. Much of this confusion has to do with the historical development in understanding photographic processes.

Resolving power of a film primarily refers to the particular response of an emulsion to a parallel-line test. The parallel-line test is the photographing of a chart with finer and finer lines, usually equal spacing between lines and sectors of smaller lines. One, then, notes the size of lines that are still separable and above some gray level. Emulsions are generally specified in terms of the number of lines per millimeter or equivalent terms. General concepts related to resolving power and its measurement are defined by industrial standards.[†]

Other terms that affect the image structure characteristics are graininess, granularity, sharpness, acutance, turbidity, contrast, adjacency effects, and Mackie lines. For more details on these terms, the reader is referred to the work of Higgins and Jones [46] and others [45, 47].

Mathematical Model for Photographic Material

The description of the use of photographic materials in optical systems requires that a mathematical model be developed which represents the response of this system element. In the linear region of the Hurter–Driffield (H–D) curve (Fig. 6-1), an expression relating the density and log exposure can be written as

$$D = \gamma_n \log E - D_0 \tag{6-4}$$

where γ_n is the slope of the linear region of the curve, and D_0 is an extrapolated ordinate intercept.

The density, however, does not represent the role of the film in a system. Therefore, an intensity transmittance $T(x, y)$ is defined as a ratio of the transmitted intensity $I_t(x, y)$ and the incident intensity $I_0(x, y)$. The trans-

[†] "American National Standard Method for Determining the Resolving Power of Photographic Materials," PH 2.33—1969, ANSI, Inc., 1430 Broadway, New York, New York 10018.

mitted intensity is a spatially averaged quantity, averaged over a small neighborhood of the point (x, y); however, averaging is done over an area large compared to grain size. This transmittance is related to density through the definition presented in Eq. (6-1). Using Eqs. (6-1) and (6-3) in Eq. (6-4), an expression for intensity transmission of the negative is obtained,

$$T_{\mathrm{n}} = C_{\mathrm{n}} I_{\mathrm{i}}^{-\gamma_n} \tag{6-5a}$$

where

$$C_{\mathrm{n}} = 10^{D_0} t_{\mathrm{i}}^{-\gamma_n} \tag{6-5b}$$

In an incoherent system, it is sufficient to examine only the intensity that photographic materials transmit. In general, a power-law relationship is used to describe intensity transmittance.

The description is made more complicated by the fact that the recording of the original intensity results in opacity where there was brightness in the input scene. This first film recording is thus known as a negative. To obtain a transparency which transmits the original scene as photographed, a positive has to be made by exposing another piece of film with light of uniform intensity transmitted through the negative. This new film is called a positive. Part of the problem in writing out a mathematical expression for this two-step process is that there are so many different intensities present. There are the initial intensity $I_{\mathrm{i}}(x, y)$ and the uniform intensity I_0 used in illuminating the negative. In addition, there are the two gammas, γ_1 and γ_2, associated with each step. Bringing this all together, the positive transmission function $T_{\mathrm{p}}(x, y)$ can be written,

$$T_{\mathrm{p}}(x, y) = C_{\mathrm{p}}(I_0 T_{\mathrm{n}})^{-\gamma_2} \tag{6-6}$$

which, when Eq. (6-5) is substituted, results in

$$T_{\mathrm{p}}(x, y) = C_{\mathrm{p}}' I_{\mathrm{i}}^{\gamma_1 \gamma_2}(x, y) \tag{6-7a}$$

where

$$C_{\mathrm{p}}' = C_{\mathrm{p}} C_{\mathrm{n}}^{-\gamma_2} I_0^{-\gamma_2} \tag{6-7b}$$

Thus, even though a two-step process is used, a positive power-law relationship results. It then becomes advantageous to attempt to control γ for various films, which is done in part through the control of development time as shown in the Fig. 6-3.

Control of the magnitude and sign of γ in Eqs. (6-5) and (6-7) results in system elements that can be called inverters and multipliers. Inversion is

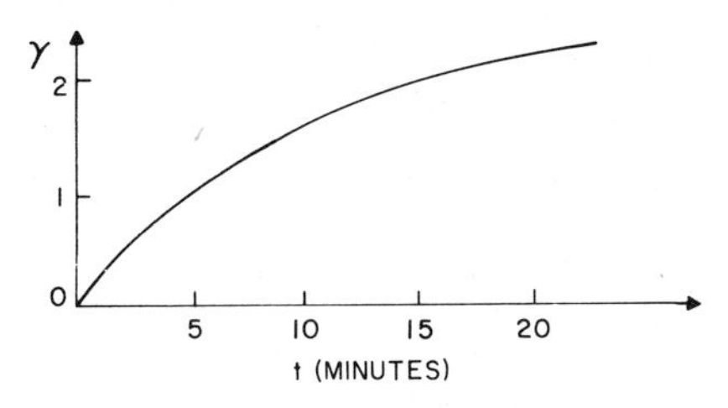

Fig. 6-3. Variation in gamma, γ, for different development times in minutes.

achieved by simply using a γ of 1. The intensity transmittance becomes

$$T(x, y) \sim \frac{1}{I_{\mathrm{i}}(x, y)} \tag{6-8}$$

An equivalent gamma of minus one can be obtained in the two-step process to give a simple multiplier, i.e.,

$$T(x, y) = I_{\mathrm{i}}(x, y) \,|_{\gamma_1 \gamma_2 = 1} \tag{6-9}$$

The idea of using various gammas will become even more important later when amplitude transmission functions for various operations are developed.

Coherent Transmission Functions

The use of photographic film as a filter system in a coherent processor necessitates the development of a corresponding amplitude and phase transmission function. The phase portion is the hardest part to generate. This is particularly true since both the emulsion and the backing material (glass or acetate) have nonuniform thicknesses. In the coherent system, these nonuniformities cause an undesirable random phase modulation. From the pragmatic point of view, any undesired phase modulation can be suppressed by mounting the filter (emulsion plus backing) in a liquid film gate. This film gate is composed of a liquid medium whose index of refraction closely matches the filter material. This matching medium is held between two optical flats as shown in Fig. 6-4. As can be seen in the figure, the liquid essentially causes the emulsion and backing to appear as flat phase surfaces. The net effect is to produce only spatial amplitude modulation on the transmitted fields. This technique of suppressing undesired phase modulation makes it possible to represent an amplitude transmission

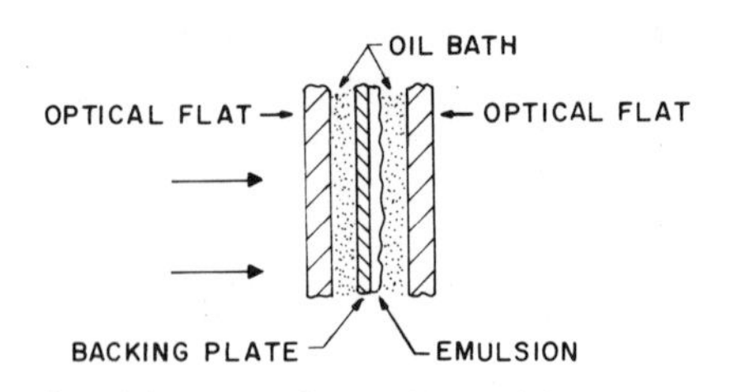

Fig. 6-4. Film gate for phase corrections. The emulsion and its backing are immersed in a bath of index-matching oil.

function as the positive square root of the intensity transmission function,

$$t_a(x, y) = [T(x, y)]^{1/2} = CI^{\gamma_1\gamma_2/2}(x, y) \tag{6-10}$$

where $\gamma_1\gamma_2$ can be plus or minus depending on whether a one-step negative or two-step positive process is used. This result has an added feature of providing a means for performing the square-root ($\gamma_1\gamma_2 = 1$) operation on a system function.

Synthesis of Quotients of Transmission Functions

Suppose it is desired to synthesize a filter with an amplitude transmission function

$$t_a(x, y) = \frac{N(x, y)}{D(x, y)} \tag{6-11}$$

Furthermore, suppose that $N(x, y)$ is to be proportional to $U_1(x, y)U_1^*(x, y)$ and $D(x, y)$ is to be proportional to $U_2(x, y)U_2^*(x, y)$. In order to build the numerator $N(x, y)$, a film is exposed and developed using a gamma of plus one. This gives an amplitude transmission coefficient of

$$t_{a_1} = C_1(U_1U_1^*)^{-1/2} \tag{6-12}$$

This filter is now used as a mask in a system like that shown in Fig. 6-5. The field U_2 incident onto an unexposed film is the product of a uniform

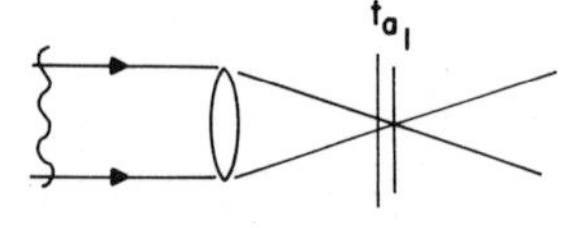

Fig. 6-5. Cascading of filters to build quotient filters. Filter t_a modulates the amplitude incident onto the second mask film.

plane wave and the mask t_{a_1}. The net illuminating intensity for the new mask is $(U_2 t_{a_1})(U_2 t_{a_1})^*$. Thus, the new mask amplitude transmission function can be written as

$$t_{a_2} = C_2(U_2 U_2^* t_{a_1} t_{a_1})^{-\gamma_2/2} = C_2 C_1^{\gamma_2}[U_2 U_2^*(U_1 U_1^*)^{-1}]^{-\gamma_2/2} \qquad (6\text{-}13)$$

Choosing the second gamma γ_2 as plus two yields

$$t_{a_2} = C\left(\frac{U_2 U_2^*}{U_1 U_1^*}\right)^{-2/2} = C\left(\frac{U_1 U_1^*}{U_2 U_2^*}\right) = C_3 \frac{N(x, y)}{D(x, y)} \qquad (6\text{-}14)$$

Thus by careful choice of gammas in the cascaded system, filters having prescribed ratios can be built. Both the numerator and the denominator have known spatial variations.

Modulation Transfer Function

The concept of transfer function was developed in Chapter IV for general optical systems. This concept is based on the linearity property of optical systems. It would seem that this linearity fails for photographic filters since the H–D representation was shown to be nonlinear. However, this dilemma can be resolved by restricting operation to the straight-line portion of the H–D curve, and then modeling the film as a linear element cascaded with a constant nonlinear element.

Referring to Fig. 6-6, the one-dimensional spatial variation can be

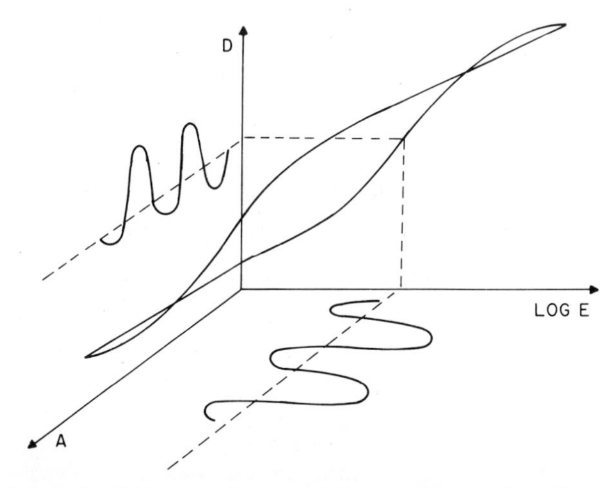

Fig. 6-6. H–D curve for a one-dimensional variation along A.

described along one axis A and the H–D parameters along the other two. If the film were exposed to a sinusoidal grating, some grating profile would appear in the density profile of the film. Measurement of the developed density profile will infer through the H–D curve what the input grating was like. This inferred exposure will, then, be the hypothetical input. A ratio can be developed using this inferred input and the known input. This ratio is called the modulation transfer function [48].

Specifically, the input exposure profile† is represented by

$$E_i(x) = E_o + E_1 \cos vx \tag{6-15}$$

where v is related to spatial frequency and E_0 is the nonlinear offset constant. With this input, an effective exposure is inferred,

$$E_{eff} = E_o' + E_1' \cos vx \tag{6-16}$$

In this case, E_1' may be complex, since there may be different phase shifts for different spatial frequencies. That is, edges may not appear as sharp edges.

The input can be represented as having a certain depth of modulation defined by the ratio of the grating peak-to-peak variation to that of the offset exposure. This defines $M_i(v)$,

$$M_i(v) = E_1/E_o \tag{6-17}$$

Similarly, $M_{eff}(v)$ is defined

$$M_{eff}(v) = E_1'/E_o' \tag{6-18}$$

The modulation transfer function is then defined as

$$M(v) = M_{eff}(v)/M_i(v) \tag{6-19}$$

Since E_1' may be complex due to some spatial phase variation with frequency, $M(v)$ can be written as having amplitude and phase terms,

$$M(v) = \mathscr{M}(v)e^{-jm(v)} \tag{6-20}$$

The term $\mathscr{M}(v)$ is then the amplitude of the modulation transfer function, and $m(v)$ is the phase term. If $m(v)$ and $\mathscr{M}(v)$ are incorporated in Eq. (6-16)

† v has the dimensions of inverse length and will be related to spatial frequency (cycles/mm). Normally, film cutoff frequencies are associated with resolving power given in line pairs per millimeter.

by using a complex exponential representation, a similar cosine representation can be obtained by taking the real part, leaving

$$E_{\text{eff}} = E_0' + \mathscr{M}(v)E_1 \cos[vx - m(v)] \tag{6-21}$$

Equation (6-21) is an important equation for representing the film element in our system. Generally it can be described as a low-pass filter. The cutoff frequencies (cycles/mm) vary widely with various choices of film.

It should be noted that the MTF is a function of spatial frequency (cycles/mm) because it is by definition the response due to sinusoidal inputs. Film, however, is generally described by resolving power based on bar chart tests (lines/mm). Although it is reasonable to infer good resolving power from an MTF with a high spatial frequency cutoff, the inference is not conclusive because of the added dependence of resolving power on granularity and microsensitometry. Modulation transfer function is a concept that fits well with the system design approach. Figure 6-7 shows some hypothetical MTF curves for common film. The resolving powers associated with these films are typically 50 lines/mm for Tri-X and 2000 lines/mm for spectroscopic Agfa 10E75.

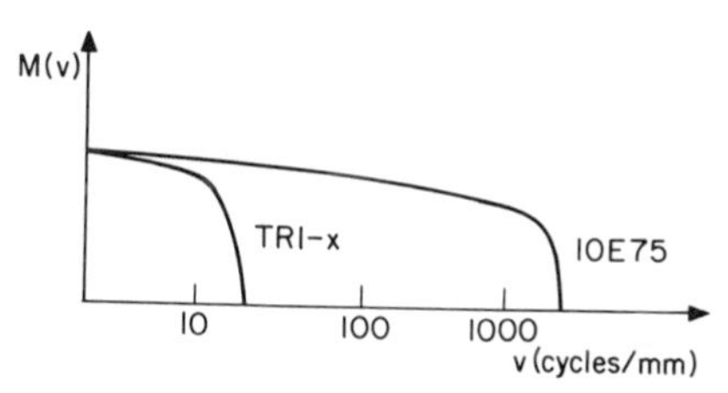

Fig. 6-7. Modulation transfer function for two types of film: high resolution and low resolution.

Television

So far in the development of interface devices, the recording of a particular intensity profile in a plane has been described in terms of density of silver. Considerable detail has been presented on the intensity detection process and how it might be modeled mathematically. There are, however, other ways of detecting and recording the intensity distributions that exist in a given plane. Devices that convert the intensity at a point or a collection of points into a time signal is the subject of this section [49].

The most common device of this type is the television system. Basically, this system converts the intensity at $I(x, y)$ of an amplitude distribution

$g(x, y)$ into a time-sequenced electronic signal. The voltage at a particular time in the sequence is proportional to the intensity of the signal at a particular point. In order that the $|g(x, y)|^2$ function is reproduced on some other output tube, the electron beam of the receiver picture tube and scanning system of TV camera are synchronized together such that input and output points roughly correspond. These synchronizing pulses are transmitted with the video information. Basically, the electronic signal is derived by having a photoactive material which emits electrons due to an incident intensity distribution. These emitted electrons are collected in the TV camera tube (vidicon or image orthocon) by an electronic scanning system. This system scans the emitting array, and where electrons are being emitted, the output current of the TV tube will change accordingly. This synchronized output current is transmitted by various means to receivers that simply modulate an electron beam that is bombarding a photoactive material. A photoactive material emits light wherever electrons cause an interaction with the tube surface. Thus the $|g(x, y)|^2$ distribution is recreated on a surface, and in times on the order of milliseconds, pictures or images are observed.

Admittedly, a lot of details on scanning, synchronizing, raster formation, flicker rates, etc., have been omitted from this discussion. The basis for this overview is to emphasize that TV systems have a role in processors even when these devices are only thought of as optical to electronic and electronic to optical converters. Television is very useful in aligning optical systems, checking outputs before photographic work begins, and even in providing the option of some electronic signal processing while the information is in electronic form. There are additional advantages. One such case occurs in correlators which rely on the inherent time integration of the viewing process. In this case, the video presentation provides the quasi-real-time output, which for high scan rates the eye will do the time integration or correlation. Some incoherent correlators have been built using this process [50, 51].

Photodiodes

Television is basically conceived of as an entire system. The details of any particular element are not so essential to this discussion. There are, however, a broad class of devices or system elements that are characterized by conversion of optical energy or photons to a change in electrical characteristics of the device. These detectors are useful in optical-processing systems for point-by-point measurement or for total integration across a plane. Basically these devices can be classed into types such as photoemissive, photoconductive, photovoltaic, photoelectromagnetic, quantum amplifier, or photoparamagnetic [49]. The photovoltaic devices are primarily characterized as active devices and can be used to read or monitor intensities at a point or plane without external biasing or supply systems. In some remote or extended-use situations, this property can be very useful. The photoconductive type can be viewed as a variable impedance element in which the electrical resistance characteristic decreases with an increasing number of photons. Basically the process is one of charge carrier production due to photon absorption. The photon absorption is followed by a subsequent change in energy level of the charge carriers. Photoconductive devices generally have the property that their spectral sensitivity extends quite far into the infrared when compared to other photoactive devices. The photoemissive devices are dependent on the photoelectric effect. That is, the photon energy $h\nu$ must be greater than the work function w of the material for electrons to be emitted. Note that increasing the number of photons or increasing the intensity does not produce electrons unless $h\nu > w$. In addition, it is important to realize that not all the photons generate electrons. There is a production or quantum efficiency about this process. Generally this quantum efficiency will run between 10^{-5} and 10^{-1} for photoemissive devices. Photoemissive devices have a spectral sensitivity that is generally limited to specific spectral ranges determined primarily by the energy gap characteristics of the material. One of the more common materials is silicon, which has a spectral sensitivity from about 450 nm to about 1100 nm. This bandwidth characteristic extends from the blue to the near infrared. It is not uniform over this band, but tends to be more uniform over the green to deep red as shown in Fig. 6-8.

Another characteristic of photodiodes, which may be important in some applications, is that there is a general loss of spectral information after detection and a loss of all phase information about the field. The detector itself does not distinguish between radiation at one wavelength as compared to another so long as both are within the spectral characteristics of the device. Similarly, the phase information is lost in detection because all of these devices respond to the spatial intensity or incident energy density. Basically the output is proportional to the amplitude squared of the field. Detection of the phase information generally requires some external device, such as an optical filter or interferometer. Finally, photodiodes have the desirable property of having a large dynamic change. Generally, this is much larger than the more conventional photomultiplier devices. The

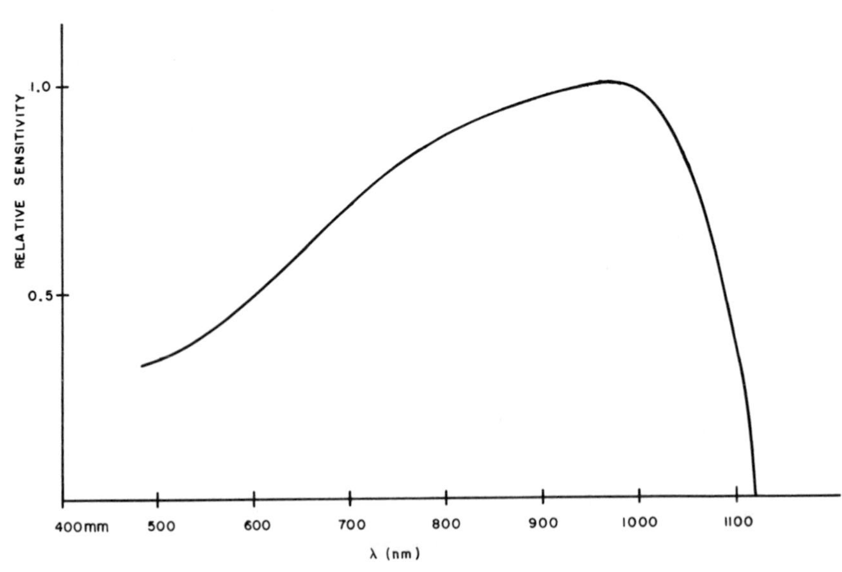

Fig. 6-8. Spectral characteristics of silicon.

trade-off parameter has generally much less postdetection gain in photodiodes vice photomultipliers; however, developing technology is slowly eroding this photomultiplier advantage, particularly in some of the PIN hybrid devices.

Photomultipliers

The photomultiplier is a photoemissive device that should be commented on specifically because it has been the workhorse of optical detection for many years. This is particularly true for detection systems where very good sensitivity is required. The photomultiplier is characterized by its high postdetection gain (10^5–10^7). It is, however, limited in dynamic range to something less than 60 dB. In addition, the lower bound on sensitivity is effected by the dark-current characteristic of these devices. This is the current that flows independent of the incident photons. The high postdetection gain is achieved by secondary emission amplification of the photoemitted electrons. This is accomplished in a series of dynodes that follow the photocathode. Many tubes have anywhere from nine to fourteen dynodes that have interdynode voltages of 200 to 400 V applied. Incident electrons on a dynode cause a secondary emission of electrons with an increase ratio of from four to ten. Thus a sequence of several dynodes can have considerable amplification. The details of such devices and the typical design characteristics are covered elsewhere [49]. One additional characteristic of photomultipliers is their large active area. Some devices have an active area of on the order of 10 to 15 in.^{2} (60 to 100 cm^2). This broad area is useful in many processing applications where it is desired to integrate spatially over the entire output plane.

Real-Time Materials

New materials have been developed that may well revolutionize the input–output capability of optical processors [52]. Two properties of these new materials make them particularly attractive. One is that with these devices both the phase and the amplitude of the transmitted fields can be controlled or modified in such a system element. Second, these phase and amplitude characteristics can be modified in near real time by external beams. There has long been a need for photochromic information storage systems that can be rapidly read and erased both in a nondestructive as well as destructive mode. Such materials need to be nearly fatigue free, have negligible thermal fade, and be insensitive to environmental parameters such as pressure, temperature, moisture, etc. One such material recently developed [52] is the alkali halide photodichroic crystals developed at the Naval Research Laboratories. The first photodichroic crystals that exhibited the best overall properties were the alkali halide crystals KCl and NaF.

Basically a photodichroic is a crystal structure where crystal defects are converted to color centers (F, M, F_A, M_A, etc.) by deposition of an electron in a vacant lattice site [53]. If a single defect is filled with an electron, this is referred to as an F center. If two F centers aggregate, this becomes an M center. The characteristic of these color centers is that they will absorb light of a certain wavelength and polarization. The M centers have the added property that at one wavelength, λ_w, and polarization perpendicular to the major axis of the color center, the radiation is absorbed. The color center will then switch its direction to the next orthogonal lattice site. In the fcc[†] structure shown in Fig. 6-9, the switched M center will be along the incident polarization. In addition, there is a second wavelength, λ_R, and polarization parallel to the major axis, at which there is absorption but no switching. The second wavelength λ_R tends to be about 1.6 λ_w.

[†] fcc is the nomenclature for face-centered-cubic crystal structure.

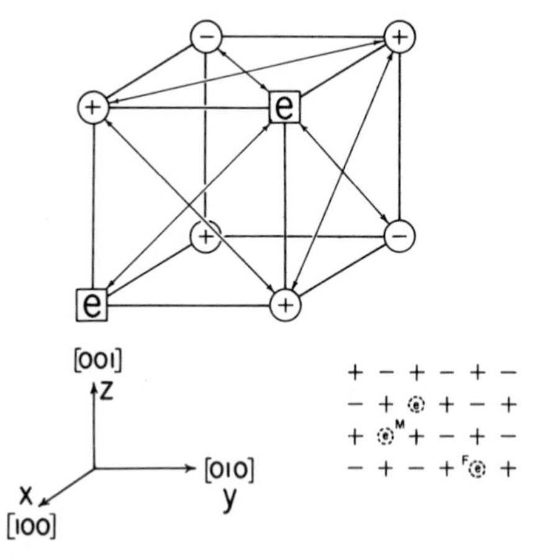

Fig. 6-9. M center model in fcc crystal lattice. The six possible figure axes are shown by arrows. Possible M center and F center defects are shown in the valence diagram.

This dual characteristic makes these crystal structures have the desirable property that at some frequencies there is a read–write–erase capability, while at another frequency there is a read-only capability. Thus it is possible to conceive of systems in which simultaneous reading (read only) and writing can occur. This is particularly valuable in the case of dynamic library-type storage systems, and in systems where iterative-type calculations are done, or where successive solutions need to be continuously generated.

Since the range of optical density depends on the density of color centers, it is important to note that densities on the order of 10^{18} to 10^{20} centers/cm^3 have been reported. These kinds of densities have led to opacities on the order of 14 o.d. in some thicknesses. This exceeds classical photographic film systems by many orders of magnitude.

These new materials have been described primarily in terms of their absorption properties. However, from the classical Kramer–Kronig susceptibility relations [54, 55], it is known that there is also an accompanying dispersion. The absorption and dispersion can be related mathematically through the susceptibility $\chi(\omega)$, associated with the classical permittivity ε, as

$$\chi(\omega_0) = \chi'(\omega_0) - j\chi''(\omega_0) \tag{6-22}$$

The real and imaginary parts, $\chi'(\omega_0)$ and $\chi''(\omega_0)$, are related by

$$\chi'(\omega_0) = -\frac{2}{\pi} P \int_0^\infty \frac{\omega\chi''(\omega)}{\omega^2 - {\omega_0}^2} \, d\omega \tag{6-23}$$

and

$$\chi''(\omega_0) = \frac{2}{\pi}\,\omega_0 P\int_0^{\infty}\frac{\chi'(\omega)}{\omega^2-\omega_0^{\,2}}\,d\omega \tag{6-24}$$

where P represents the Cauchy principal part of the integral. The systems are considered to be linear, stationary, and causal. The physical proof of these constrained interrelationships was recently demonstrated [56].

Equations (6-23) and (6-24) can be reduced for the kinds of optical systems under consideration. The simpler relations

$$\Delta\eta = -\{A\}\,\frac{2\Delta\nu(2\nu_0+\Delta\nu)}{\Delta\nu^2(2\nu_0+\Delta\nu)^2+\Gamma^2(\nu_0+\Delta\nu)^2} \tag{6-25}$$

and

$$k = \{A\}\,\frac{\Gamma(\nu_0+\Delta\nu)}{\Delta\nu^2(2\nu_0+\Delta\nu)^2+\Gamma^2(\nu_0+\Delta\nu)} \tag{6-26}$$

where

$$\{A\} = \frac{N_0 f e^2}{8\pi m}\,\frac{(\eta_0^{\,2}+2)^2}{\eta_0} \tag{6-27}$$

are useful and lead to the index of refraction N, which is defined as

$$N = \eta - ik \tag{6-28}$$

The constants above are Γ, the dampening constant; ν_0, the center frequency; and f, the oscillator strength. Also inherent in the above is the assumption of the Lorentz relation

$$\frac{N^2-1}{N^2-2} = \frac{4\pi}{3}\,N_0\alpha \tag{6-29}$$

where α is the polarizability

$$\alpha = \left(\frac{e^2 f}{4\pi^2 m}\right)\frac{1}{\nu_0^{\,2}-\nu^2+i\Gamma\nu} \tag{6-30}$$

An elementary plot showing these coupled effects is presented in Fig. 6-10. It is quite important to note that measurable dispersion effects exist over much broader bands than do the corresponding absorption effects. The net result is that M band dispersion can be controlled by F band light with seemingly little absorption. The order of magnitude of this effect on $\Delta\eta$ is approximately 10^{-2} for a color center density of about 10^{18} centers/cm^3. These effects may seem small, but the possibility of simultaneous phase and amplitude modulation makes these materials particularly attractive for complex filter applications.

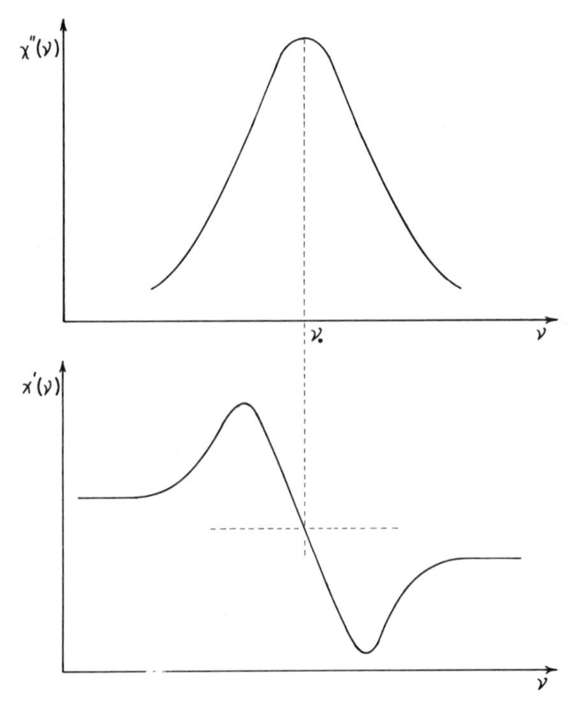

Fig. 6-10. Absorption $\chi''(\nu)$ and dispersion $\chi'(\nu)$ spectra of M centers.

Real-Time Materials Applied to Integrated Optics

Recent work [57] has demonstrated that thin-film optical structures can have bulklike properties capable of supporting guided waves [9]. Generally, the necessary condition for wave guiding is simply derived from Snell's law. For a simple thin-film layer it is necessary that the refractive index of the layer n_l be greater than the refractive index n_b of the substrate boundary. The possibility of creating photodichroic structures in thin films suggests that it may be possible to fabricate real-time, modifiable integrated optical circuits and systems that are on the order of a few millimeters in total size. Propagation of guided modes in alkali halide thin-film structures has been demonstrated [17]. With the possibility that similar modes will propagate in a photodichroic alkali halide thin-film structure, the potential for real-time, modifiable optical elements in integrated optics is very real.

The control of both the amplitude and the phase through remotely controlled color center changes provides the basis for generalized processor elements. It is important to note that the amplitude and phase would not be completely independent. The interrelationship is governed by the Kramer–Kronig relations [55] as shown in Eqs. (6-23) and (6-24). With these kinds of processor elements, it would be possible to write, erase, and read the control functions in near real time.

Since this control is effected by external beams, added degrees of freedom are provided. This is particularly the case in bulk storage devices where the third dimension of access can be achieved through the infrared switching immobilization [52]. Further, it is possible to create small changes in the local dielectric constant of the layer. This is achieved by changing the real part of the susceptibility by external beams. Thus, light channels, fiber optic multiplexing switches, gratings, interferometers, or filters can be created as processor elements.

With the demonstration of F_A color center laser action [58], gain in the alkali halide integrated optical processor now seems possible. A conceptual processor can be conceived using the elements described above. One futuristic possibility is the processor system shown in Fig. 6-11, which is essentially a complete processor on a chip. Many other possibilities can and will be envisioned [9, 57], particularly as data rates in data processors exceed the apparent hundred-megabit-per-second limit of large-scale integrated digital processors.

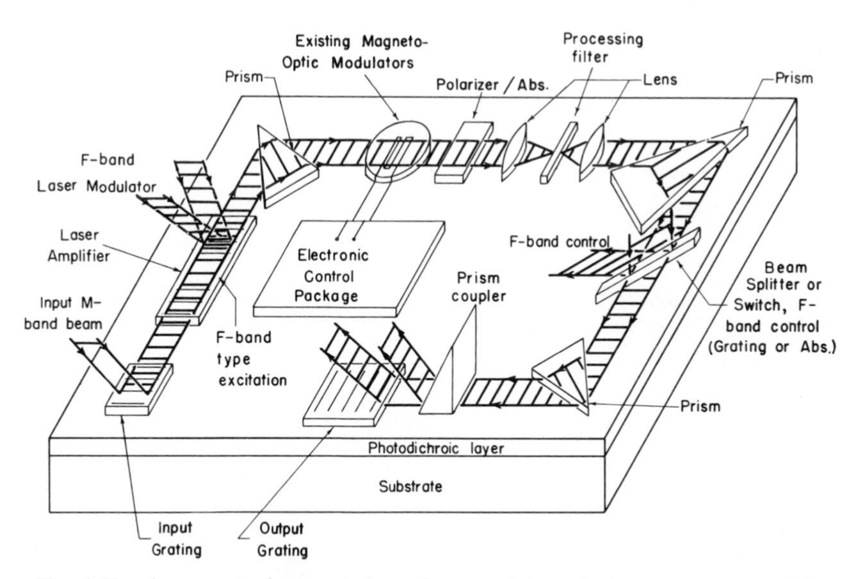

Fig. 6-11. A conceptual presentation of a complete optical processor on a chip.

Problems

1. Given the curve of Fig. 6-1, what would be recorded on the film regarding an image, $g(x, y)$, if the average exposure were 0.002 m-candle sec?

2. Calculate the exposure for an image that depends on time as follows:

$$I(t) = I_0 \exp\left[\frac{-(t-\tau)^2}{T_0^2}\right]$$

where $T_e \gg \tau, T_0$.

3. What dynamic range of photodichroics can be expected in terms of optical density units, o.d.?

4. What would happen to object detail if a photograph were taken in sequences, on the same film, such that the shoulder of the curve is exceeded?

5. Determine the process and γ's required to make an amplitude transmission characteristic of

$$t_a = \frac{|U_2(x, y)|}{|U_1(x, y)|}$$

Be sure to list how many steps are required and what γ is used in each step.

6. The classical model of density deposit D in film due to an exposure E is the so-called linear model which relates D and E through the γ. Assume that one would like to use the knee of the curve for a nonlinear relationship between D and E. Derive the appropriate equations for a parabolic fit.

Chapter VII

INTERFEROMETRY

Introduction

In Chapters I and II the description of fields in a removed aperture was given in terms of the excitation in a source aperture by an integral equation. The approximations necessary to treat this problem in terms of rays or beams of light were given in the section on geometrical optics. The intensity of the light at the receiving aperture in such systems was given by means of the square of the amplitude in the wave picture. This allowed a simple picture for describing intensity variations. However, when two different ray bundles are superposed, assuming that they are of the same wavelength and polarization, the resulting intensity distribution is not a simple superposition of the results obtained by considering each ray bundle alone. The description of the intensity distribution of this new system has to account for the phase length associated with the path that each ray bundle traverses. These phase differences cause the two bundles to interact in both constructive as well as destructive ways. The resulting interference patterns are useful in determining properties of the original beams and constitute an important tool in the measurement of the phase of optical waves.

The light and dark variations in intensity that appear to modulate output patterns are called fringes. The number of fringes, their spacing, and their relative amplitudes are the parameters that are associated with the interferring process and the basic properties of the original exciting sources. The amount of interaction depends not only on the phase differences, but on the correlations between the two wavelengths, the polarizations, the stability of the wavelengths, polarizations, phases, and amplitudes.

Interference and its associated phenomena are observed quite commonly. If the illumination of a sharp edge is projected onto a screen, fringes are observed; if a street light is observed through a window screen, a crosslike pattern is seen; if two very smooth and nearly flat transparent objects are placed in contact, fringes are observed when monochromatic light is used to illuminate them; or if oil is spilled on water, multicolored patterns are seen. All of these are examples of interference.

Some of these observed processes have been described in Chapter II by diffraction theory in distributed apertures. This full-wave picture did not separate the phenomena, but rather treated the resulting fringe patterns as due to the array factors. However, this picture does not fit with much of the historical development or the evolution of many testing devices currently in use today in such areas as spectroscopy, meteorology, or precise machining. In addition, it is often not convenient to reduce the physical setup to a mathematical framework that fits easily into the diffraction equations. Therefore, it is useful to have a description that is consistent with the geometrical optics development and allows the conceptual understanding of these observed phenomena with the simpler tools of ray optics.

There are many ways to obtain two or more beams or ray bundles. It is convenient, however, to divide the methods and results into two groups. The first can be described by divisions in the amplitude of a wave, also referred to as a temporal comparison or axial comparison. The second is wave front division, also referred to as a transverse comparison on a phase front. It will be seen that in discussion of cavity modes, this comparison relates to the transverse mode structure.

In amplitude division the whole wave front is being split, and different wave fronts from different epochs are being compared. This can be done by a beam splitter and folding mirrors or other means to essentially fold two different time elements onto an output plane or device.

In wave front division, different portions of a given wave front are brought together and superposed. In this case the phase differences across a phase front are compared. The relative phase differences allow a description of the transverse phase variations. Grouping the division processes into these two groups enables the separation of the various kinds of interferometers and provides a clearer relation among the common physical effects.

It may appear that interference or interferometry is related only to a ray-optic picture. This constraint is not implied, since interferometric concepts are widely used in wave analysis methods. A good example was presented in Chapter V in the discussion of matched filtering. The matched filter was referred to as an interferogram. Another example will be presented in Chapter VIII on holography. In holography an interferogram is used in the reconstruction of the desired images.

There are also higher orders of interference. The classic experiments of Hanbury–Brown and Twiss [59] are excellent examples of intensity interferometry. If one considers interferometry in the simplest sense as a product of complex amplitudes, then it is referred to as a second-order phenomena. Intensity interferometry is associated with the product of intensities and can be considered a fourth-order product in amplitudes. Thus intensity interferometry is often referred to as fourth-order interference. Some current work [60–62] in intensity interferometry is associated with determining source distributions in regions where the propagating medium is corrupting the received image beyond a point where classical detection methods can be used. More is presented on intensity interferometry in Chapter IX.

The above discussion is meant to provide an overview on the use of interferometric tools or simply interferometry. The material to follow will outline some of the basic interferometers, and then lead into a discussion of the Fabry–Perot interferometer, which is used in laser cavities.

Young's Interferometer

One of the simplest demonstrations of wave front division interferometry is the classical Young's experiment [11, 59], which uses two slits or apertures in an incident beam. The resulting fringe pattern is related to slit spacing, wavelength, and the distance to the viewing screen. In Fig. 7-1 a diagram of the setup is presented. A plane wave is assumed to be incident from the left onto an absorbing screen that has two slits separated by a distance d.

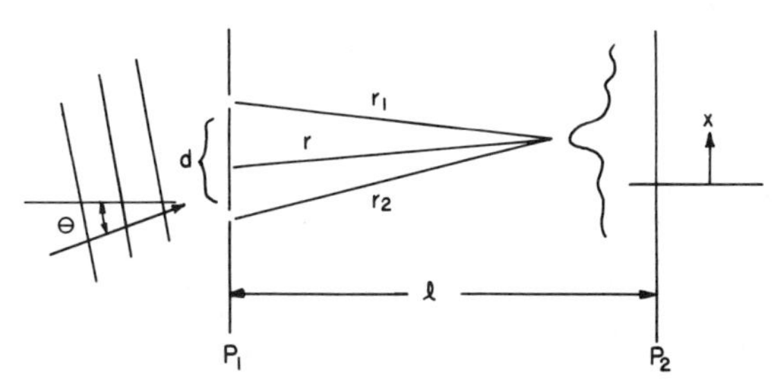

Fig. 7-1. Young's interferometer with an incident plane wave at an angle θ and slit spacing d. Fringes are formed in a plane P_2.

The physical path lengths from each slit to the point of observation are designated as r_1 and r_2, respectively. The distance from the plane of the slits P_1 to the observation plane P_2 is l. The transverse distance of the observation point from the optical axis or axis of symmetry of the system is designated by x. The plane wave is assumed to be incident at an angle θ. The effect of this angle θ is to cause the peak of the interference pattern to appear at some point on the opposite side of the axis of symmetry. In addition, the incident light is assumed to be quasi-monochromatic.†

The pattern observed on the observation screen is composed of light and dark fringes that arise from the constructive and destructive interactions of the ray bundles. Given that everything else (wavelength, amplitude, and polarization) remains the same, destructive interference will occur when the electrical phase lengths of the paths r_1 and r_2 differ by odd multiples of $\lambda/2$. Constructive interference will occur for multiples of λ. These conditions can be written [13] in terms of a distance off axis by ($l \gg d, x$)

$$x_c = m(\lambda l/d) \qquad m = 0, \pm 1, \pm 2, \ldots \tag{7-1a}$$

$$x_d = \left(\frac{2m+1}{2}\right)\frac{\lambda l}{d} \tag{7-1b}$$

Fringe spacing can be written as

$$\Delta x_c = \lambda l/d \tag{7-2a}$$

or

$$\Delta x_c = 2\pi l/\eta k_0 d \tag{7-2b}$$

where k_0 is the wave number of free space and η is the index of refraction of the medium. The position of the central maximum can be related to incident angle θ by

$$\theta = \tan^{-1}(x_0/l) \tag{7-3}$$

which can be useful in determining angles of arrival of incident waves.

Rayleigh Interferometer

One of the assumptions in the Young's interferometer is that the distance l is very large compared to the transverse distances d and x. This stringent requirement can be overcome by using a lens. In principle, the focal plane

† Quasi-monochromatic is defined by the constraint $\Delta f \ll f_0$, where $\Delta f = f(t_2) - f(t_1)$ and $f_0 = [f(t_2) + f(t_1)]/2$.

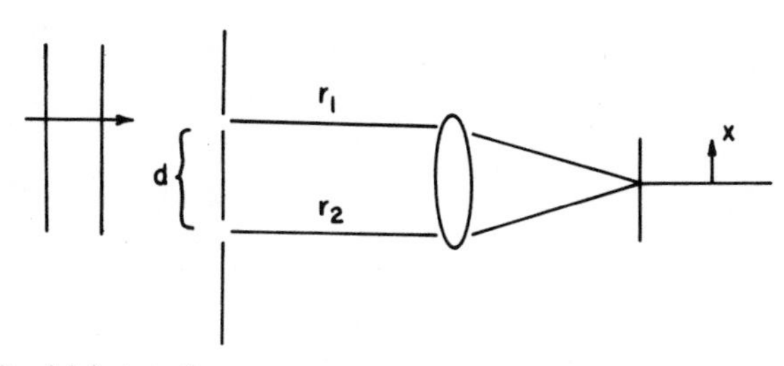

Fig. 7-2. Rayleigh interferometer with a normally incident plane wave and a lens to form an equivalent far field.

of a lens can be thought of as a point far removed from the collimated source where the rays come together. Thus the distance l is equivalently very large in the system with a lens. This system is called a Rayleigh interferometer. Such a system is shown in Fig. 7-2.

The Rayleigh interferometer is quite sensitive to the relative phase shifts in the paths r_1 and r_2. This characteristic can be used to great advantage in testing of phase objects. The fringe spacing in the output plane is the same as given by Eq. (7-2). These fringes, however, will be shifted if the electrical path length difference $(r_1 - r_2)k$ changes. The amount of fringe shift is related to the fraction of 2π radians changed during the perturbation in each path, as given by

$$\xi = \Delta r/\lambda_0 \tag{7-4}$$

Thus the amount of shift is the product of Eqs. (7-4) and (7-2).

If one path r_2 has a medium of refractive index η_2 and path r_1 has η_1, then the change in fringes is related to

$$\xi = (\eta_2 - \eta_1)L/\lambda_0 \tag{7-5}$$

where L is the length of each phase medium in the paths r_1 and r_2. This result is useful in plasma diagnostics and other refractive index measurements. It has been used [11] to measure differences in index of refraction on the order of 10^{-8}. A slight variation in the result given in Eq. (7-5) enables the application of this interferometer to lens stigmatic and aberration tests [11].

Michelson Stellar Interferometer

Another variation of the basic Young's interferometer is the Michelson stellar interferometer shown in Fig. 7-3. In this case the mirror system M_1, M_2, M_1', and M_2' is used to gain an equivalent increase in the aperture separation. This increase in basic separation is important because the shift in fringe orders is proportional to the angular separation of two sources or the angular width of an extended source, the separation of the sampling apertures (M_1 and M_1'), and the inverse of the average or effective wavelength. For small angles this can be written as

$$\xi \cong \theta d/\lambda_0 \tag{7-6}$$

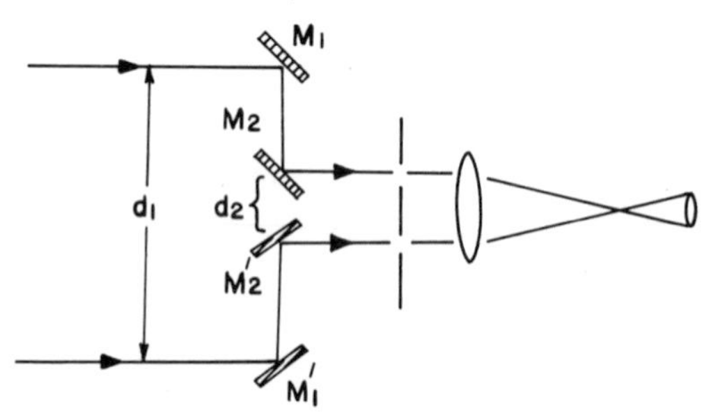

Fig. 7-3. Michelson stellar interferometer for sampling with an equivalent spacing of d_1 with mirrors M_1–M_1' that fold onto mirrors M_2–M_2' for the sampling slits spaced at d_2.

Now as the separation of the mirrors (M_1 and M_1') is changed, the superposed fringes will show succeeding bright and dark distinctness. The first minimum of distinctness will occur when

$$d = \lambda_0/2\theta \tag{7-7}$$

This condition can be used, through a measurement of d and λ_0, to determine θ, the angular separation. It has been used to measure the angular diameter of the star Betelgeuse. The value was found to be about 0.047 sec of arc.

Michelson Interferometer

The simplest interferometer of the amplitude division type is the Michelson interferometer shown in Fig. 7-4. This interferometer should not be confused with the Michelson stellar. In the case of the Michelson interferometer, the input parallel beam is divided at the beam splitter, BS. Part of the wave propagates to mirror M_1, which is movable, and part to mirror M_2. Both waves are reflected at the mirrors and return via beam splitter to be superposed at the plane P_1. Interference fringes are formed

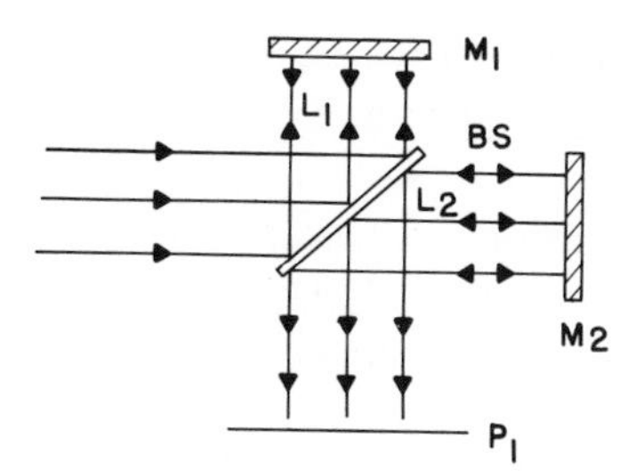

Fig. 7-4. Michelson interferometer for amplitude separation of the input wave.

whenever the path length difference $L_1 - L_2$ is an odd multiple of $\lambda/4$. Since M_1 is movable, it is convenient to think of an equivalent mirror M_2' which is adjacent to M_1. Conceptually, then the interferometer is responding to the thickness of the air film between M_1 and M_2'. If the film is wedge shaped, the fringes will be displaced and can be estimated to better than 1/20 of an order.

Although not used extensively today, the interferometer was originally used in the famous Michelson–Morley ether-drift experiment. It was also used to do the earliest work on fine structure of spectral lines. Its simplicity makes the interferometer of considerable theoretical interest, particularly in such areas as the study of partial coherence.

Twyman–Green Interferometer

If a Michelson interferometer is illuminated with collimated quasi-monochromatic light and a lens is added to the output plane as shown in Fig. 7-5, the interferometer becomes capable of measuring very small tilts in mirror M_2 with separated paths L_1 and L_2. This technique allows careful

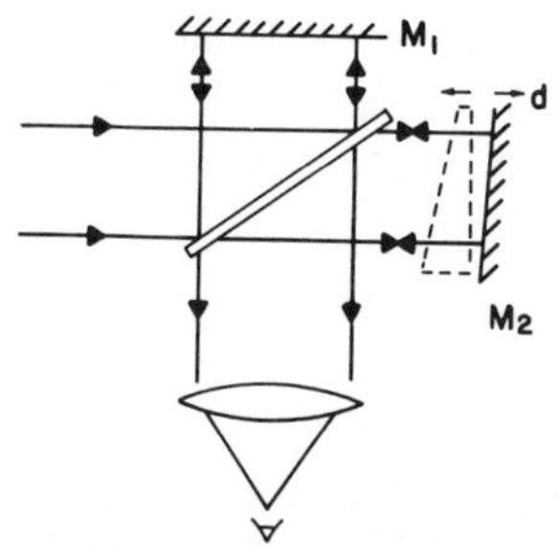

Fig. 7-5. Twyman–Green interferometer for amplitude separation where a lens is used to form an equivalent far field. Mirror M_2 is often tilted to allow prism measurements.

measurement of the parallelism and flatness of optical elements placed in leg L_2. A bright fringe will form if the electrical thickness of the element is some integer multiple of a wavelength λ_0. In general, the fringes form parallel straight lines starting at the apex of the equivalent wedge formed by M_2 and the virtual image M_1'. If M_1 is also adjustable in tilt, the amount of tilt can be measured (mechanical measurement) by noting when the fringe field of view becomes uniform.

It is easy to see that appropriate adjustment in the tilt of M_2 would allow measurements of prisms by making the surfaces of the prism appear as virtual flat surfaces. In addition, one can also consider making M_2 a spherically convex surface whose center of curvature is aligned with a lens focus. The aberrations and astigmatism of such lenses become quite pronounced. The method is then useful for lens quality measurements.

The Kösters interferometer and the Dowell interferometer are extrapolations [11] of the Twyman–Green interferometer as applied to gauge and length measurements. The Kösters interferometer is generally associated with measurements of a single gauge, whereas the Dowell is generally used to compare two different gauges. There are many other interferometers, all arising out of variations of the basic Twyman–Green concept. Examples [11] are the Jamin, the Sirks–Pringsheim, and the Dyson interferometers. Many of these various configurations have been slightly modified and used as modified Fabry–Perot configurations for mode control in laser operation.

Mach–Zehnder Interferometer

In Fig. 7-6 an interferometer with even more separated paths is shown. This is the classic Mach–Zehnder interferometer used extensively for phase measurements where large media are being probed. This interferometer has been used in such cases as wind tunnel tests, shock tube measurements, and many other cases where extended fields of view are to be examined.

Basically the interferometer compares the differences in electrical path length from BS_1-M_2-BS_2 to BS_1-M_1-BS_2. It has considerable flexibility in that multiple wavelengths can be used on the input. This is particularly advantageous in its application to multicolor holography. In general, the electrical path length over the path M_1-BS_2 is modified by the presence of some phase-perturbing media. This modified path has a virtual path superposed along the path M_2-BS_2, such that wave fronts along the two paths are compared. Any tilts present in either wave front will appear as fringes

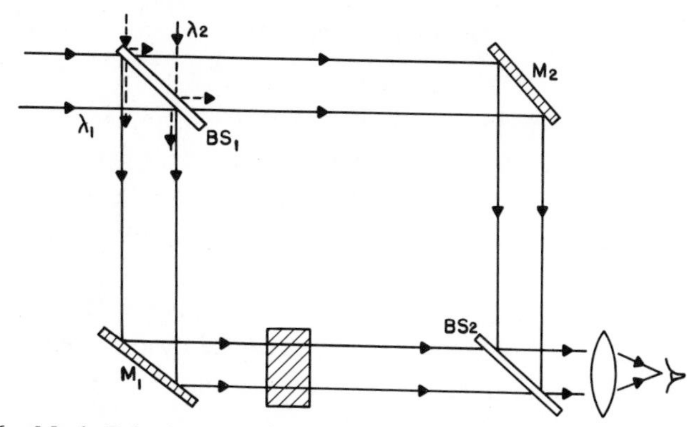

Fig. 7-6. Mach–Zehnder interferometer used extensively for phase measurements of a typical media shown as the cross-hatched box. Can be used in a two-wavelength mode to reduce density parameter influences.

in the output similar to that for the Twyman–Green interferometer. The virtual phase difference between the emergent beams is then

$$\delta = k_0 \eta \,\Delta L \tag{7-8}$$

where k_0 is the free-space wave number, η is the refractive index between the virtual wave front and the unperturbed reference wave front, and ΔL is the difference in axial position of the two wave fronts.

If the interferometer is used to test a phase media, like a plasma, then there will be a change in fringe order. This order change ξ can be calculated using Fermat's principle [Eq. (1-30)] to obtain

$$\xi = \frac{1}{\lambda_0} \int (\eta' - \eta)\, ds \tag{7-9}$$

where η' is the refractive index of the plasma and η is the refractive index of the same region without the plasma, ds is an increment along the path of the traversing ray. Inherent in this result is the assumption that the ray path has not moved significantly during the measurement.

In the multiwavelength configuration it is possible to normalize our constants pertaining to the medium and thus make it quite insensitive to undesirable density, pressure, or temperature changes. This multiwavelength implementation thus allows broader and more detailed parameter studies of media such as plasmas, shock waves, or other phase-perturbing media.

This interferometer is also used to build complex spatial filters that are used in matched filtering (Chapter V). The additional applications to holography make it a very useful and flexible interferometer.

Fizeau Interferometer

One interferometer that is similar to the Twyman–Green interferometer without all the path separation is the Fizeau interferometer. A simplified version is shown in Fig. 7-7(a). Light from the two reflections associated with P_1 and P_2 are superposed, reflected by the beam splitter BS and brought to a focus with the lens L. The fringes formed are very similar to those described for the Twyman–Green configuration. The relative tilt in the planes P_1 and P_2 will produce a tilt in the associated wave fronts which will appear as fringes from a wedge in the output plane.

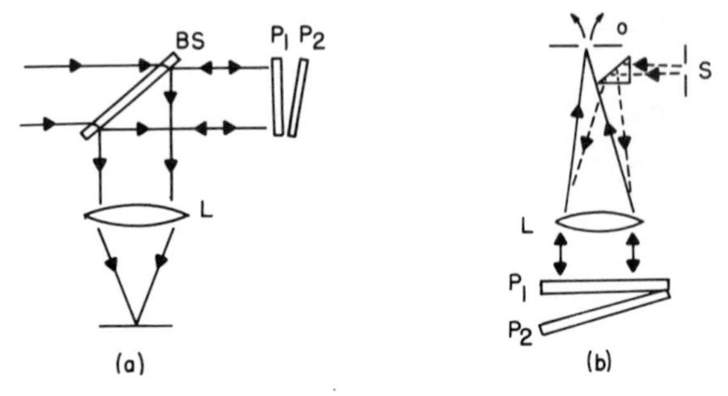

Fig. 7-7. Fizeau interferometer. (a) A simple technique for measuring the tilt in planes P_1 and P_2. (b) An alternate method not using a beam splitter.

Another version of this interferometer is shown in Fig. 7-7(b). In this case, a quasi-monochromatic source S is used to illuminate the pair of surfaces P_1 and P_2, and the fringes are observed through the aperture O. The lens serves to provide near normal incidence on the planes P_1 and P_2 and thus makes the fringes correspond to equal differences between P_1 and P_2.

One advantage of the Fizeau interferometer is that it provides a large area over which the fringes can be distinct. This condition arises from the near normal incidence of the illuminating and reflecting rays from the thin wedge-shaped film formed by P_1 and P_2. In addition, the fact that fewer mirrors are required makes the Fizeau a simpler and easier interferometer to use. Experiments with modified forms of the Fizeau interferometer have shown that it can be used to detect differences in length or separation to an accuracy of about 25 nm.

Newton Interferometer

An interferometer that can be used to display contours of equal height or to give a topological map of a flat surface is the Newton interferometer. The observed fringes have been generally referred to as Newton's rings. In Fig. 7-8, a simple setup, often used by machinists to measure flat surfaces, is shown. It is analogous to a thin-film wedge situation where the fringes are associated with contours of equal height. The bright fringes will occur for situations that satisfy

$$2\eta h \pm \lambda_0/2 = m\lambda_0 \qquad m = 0, 1, 2, \ldots \tag{7-10}$$

The extra factor of $\lambda_0/2$ occurs because of the π phase shift associated with the reflection off of the bottom surface.

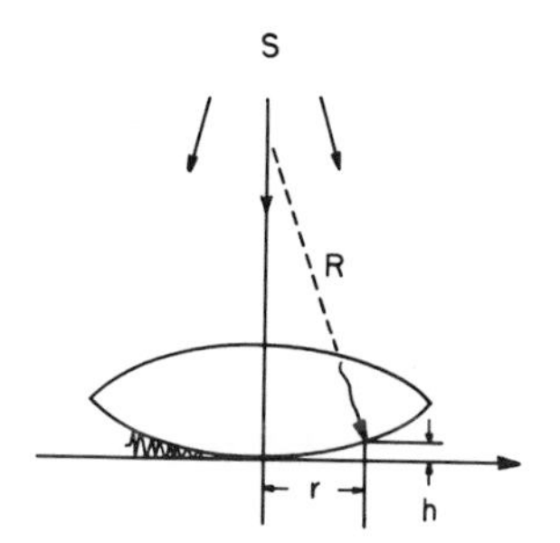

Fig. 7-8. Newton interferometer used for measuring contours of equal height.

In the case of the Newton's rings, the height h is related to the radius of curvature R of the lens and the distance from the center of symmetry to the fringe being observed, through the Pythagorean theorem

$$h = R - (R^2 - r^2)^{1/2} \tag{7-11}$$

which for $R \gg r$, reduces to

$$h \simeq r^2/2R \tag{7-12}$$

Since there is a π phase shift on reflection from the bottom, the center of this fringe pattern will be dark. Thus within the approximation associated with Eq. (7-12), the condition for a dark fringe is

$$r = (mR\lambda_0)^{1/2} \qquad m = 0, 1, 2, \ldots \tag{7-13}$$

This is an interesting result relating the fringe positions to the square root of fixed parameters R and λ_0, and the order m. Thus the fringes will appear to be compressed as one moves from the origin of this concentric system. This system, interestingly enough, could be used in a simple fashion to measure the approximate wavelength of the light being used.

In its application to surface flatness measurements, the concentric uniformity of the fringes is very important. The distortions of the fringe pattern can give a good indication of where corrections are necessary. The method, however, is more of historical interest today due to the introduction of new devices and holography.

Multiple-Beam Interference

If light is incident at an angle θ onto a glass or dielectric plate, there are multiple reflections and transmissions that occur. If the reflectance is large enough, the reflection and transmission effects appear to have very marked fringe effects. These fringes are useful in spectroscopic studies. One extension of this multiple-beam idea has led to the famous Fabry–Perot interferometer that is used extensively in laser cavities.

Consider first the simple dielectric layer of refractive index η' and thickness d shown in Fig. 7-9. A wave of specific polarization is incident at an angle θ onto the upper surface. A portion of this wave is reflected at an angle θ. The amplitude of this reflected wave will be proportional to the square root of the reflectance R. The sign of this amplitude reflection coefficient will be determined by the polarization. If η' is larger than the index of the upper region and the polarization is perpendicular to the plane of incidence, then the entrance reflection coefficient will be negative.

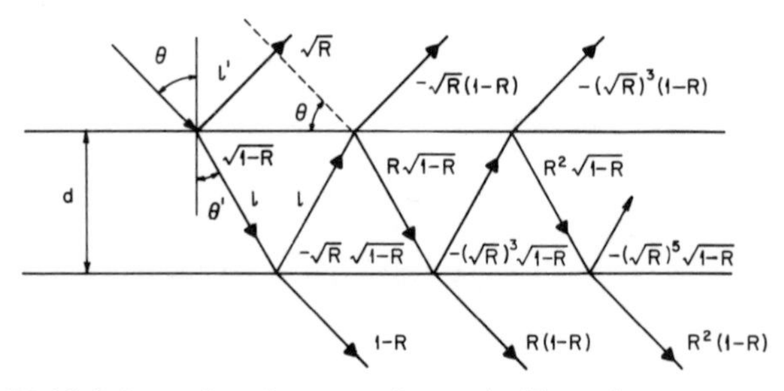

Fig. 7-9. Multiple-beam interferometer for an incident plane wave at angle θ and surface reflectivity R.

The reflection coefficient of the rays reflected internally will be positive. If the polarization is parallel, the opposite situation will occur. For purposes of this analysis, we shall assume the latter case of parallel polarization. In the final analysis, the interferometric results will not change.

A second portion of the wave is transmitted through the top surface with an amplitude transmission coefficient proportional to the square root of transmittance T or one minus the reflectance R. This follows from simple energy considerations in a lossless model, where

$$R + T = 1 \tag{7-14}$$

The transmitted wave travels through the dielectric at an angle θ' yielding to a similar transmission and reflection process at the second surface, the only difference being the sign of the amplitude reflection coefficient at the bottom.

This process continues, alternating between the upper and lower surface, with the result that the upper reflected wave and the bottom transmitted wave are really composed of many superposed waves, each differing in the amount of phase delay.

To evaluate this phase delay or phase difference for two succeeding waves, consider the following geometrical arguments. The phase difference between the first reflected ray and the second reflected ray is

$$\delta = k_0(2\eta' l - \eta_0 l') \tag{7-15}$$

where η_0 is the index of refraction of the upper medium, and η is refractive index of the lower medium. The distance l is simply related to the thickness d by

$$l = d/\cos\theta' \tag{7-16}$$

The distance l' is also related to d by simple geometry

$$l' = 2l\sin\theta'\sin\theta = 2d\tan\theta'\sin\theta \tag{7-17}$$

Substituting both Eqs. (7-17) and (7-16) into Eq. (7-15) and using Snell's law,[†] the following results for δ:

$$\delta = \frac{2k_0\, d\eta'}{\cos\theta'}(1 - \sin^2\theta') = 2k_0\eta' d\cos\theta' \tag{7-18}$$

This result will be useful later when evaluating the fringe effects.

[†] In this rotation, Snell's law is $\eta_0 \sin\theta = \eta' \sin\theta'$.

The superposition of the reflected and transmitted waves can be written in series form as

$$U_R = |U_i| \{\sqrt{R} - \sqrt{R}(1-R)e^{i\delta}(\sqrt{R})^3(1-R)e^{i2\delta} \cdots\} \quad (7\text{-}19)$$

and

$$U_T = |U_i| \{(1-R)(1 + Re^{i\delta} + R^2e^{i2\delta} + \cdots)\} \quad (7\text{-}20)$$

These series can be written in closed form for a large number of terms as

$$U_R = |U_i| \left\{\sqrt{R}\left(\frac{1-e^{i\delta}}{1-Re^{i\delta}}\right)\right\} \quad (7\text{-}21)$$

and

$$U_T = |U_i| \left\{\frac{(1-R)}{1-Re^{i\delta}}\right\} \quad (7\text{-}22)$$

The intensity of the reflected and transmitted waves can be obtained by multiplying each equation by its complex conjugate. This yields

$$I_r = I_i\left[\frac{4R\sin^2\delta}{(1-R)^2 + 4R\sin^2\delta/2}\right] \quad (7\text{-}23)$$

and

$$I_t = I_i\left[\frac{1}{1 + [4R/(1-R)^2]\sin^2\delta/2}\right] \quad (7\text{-}24)$$

which clearly have corresponding maxima and minima that vary with the phase difference δ.

The transmitted intensity reaches a maximum value equal to the incident intensity when

$$\delta = 2m\pi \qquad m = 0, 1, 2, \ldots \quad (7\text{-}25)$$

and a minimum value of

$$I_t = I_i\left(\frac{1-R}{1+R}\right)^2 \quad (7\text{-}26)$$

for

$$\delta = (2m+1)\pi \qquad m = 0, 1, 2, \ldots \quad (7\text{-}27)$$

The amount of phase change required to go from a maximum to a minimum is a π shift in δ. However, a more meaningful measure is the amount of phase shift required to go to one-half the intensity of the maximum. This occurs for

$$\sin(\delta_h/2) = (1-R)/2\sqrt{R} \quad (7\text{-}28)$$

which for reflectances R near unity reduces to

$$\delta_h \cong (1 - R)/\sqrt{R} \tag{7-29}$$

If we consider the fringe to have a width $2\delta_h$ associated with this one-half intensity level, then the finesse $\mathscr{F}$ of the interferometer can be defined as the ratio of the phase shift from maxima to maxima divided by the phase width of the half-intensity level. Thus the finesse becomes

$$\mathscr{F} = \pi\sqrt{R}/(1 - R) \tag{7-30}$$

Using this result, the reflected and transmitted intensities can be written as

$$I_r = I_i\left[\frac{\mathscr{F}^2 \sin \delta}{(\pi/2)^2 + \mathscr{F}^2 \sin^2 \delta/2}\right] \tag{7-31}$$

and

$$I_t = I_i\left[\frac{1}{1 + (2\mathscr{F}/\pi)^2 \sin^2 \delta/2}\right] \tag{7-32}$$

It should be noted that as the reflectance R gets very close to unity, the finesse gets very large. In this case I_r approaches I_i, and the transmitted intensity goes to zero except at those cases where $\sin^2 \delta/2$ goes through zero. A plot of I_t/I_i is shown in Fig. 7-10. Thus it is clear how such bright fringe lines can occur in the transmitted field of view and such dark lines can occur in the reflected pattern. The transmitted pattern is much easier to see.

The concept of resolving power can also be defined in terms of finesse and the phase difference δ' associated with the geometrical mean of the

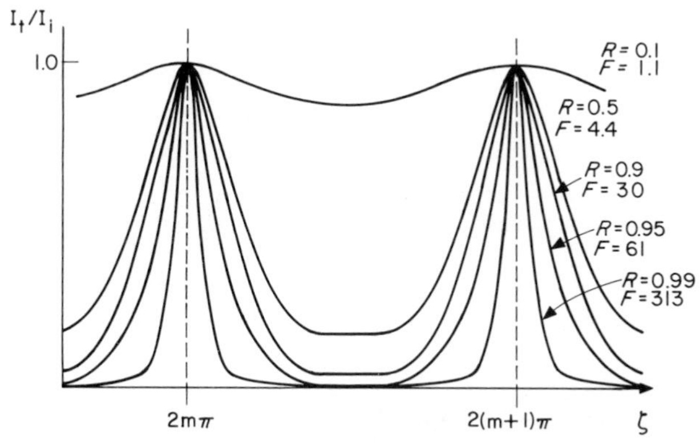

Fig. 7-10. Transmitted intensity normalized by incident intensity for a parallel plate multibeam interferometer for various reflectances $\mathscr{R}$ and finesse $\mathscr{F}$.

wavelengths at each end of the half-intensity level. It is assumed that the required phase shift to achieve the half-intensity level is derived from varying the wavelength. Thus starting with the resolving power defined as

$$\mathscr{R} = \lambda'/\Delta\lambda \tag{7-33}$$

It is necessary to write $\Delta\lambda$ or Δk in terms of the phase shift δ_h. That is,

$$\delta_h = 2\Delta k\eta' d\cos\theta' = (1 - R)/\sqrt{R} \tag{7-34}$$

which permits Δk to be expressed in terms of the finesse $\mathscr{F}$ as

$$\Delta k = \frac{1}{(\pi\mathscr{F})(2\eta' d\cos\theta')} \tag{7-35}$$

In a magnitude sense Δk is proportional to $\Delta\lambda$ as

$$|\Delta k| = 2\pi(\Delta\lambda/\lambda'^2) = (\Delta\lambda/\lambda')k' \tag{7-36}$$

Rearranging and dividing through by k', the resolving power $\mathscr{R}$ becomes equal to

$$\mathscr{R} = \pi\mathscr{F}\delta' \tag{7-37}$$

where δ' is the phase associated with the geometrical mean wavelength of the half-intensity level, which from Eq. (7-25) becomes $2m\pi$ $(m \neq 0)$. Thus $\mathscr{R}$ can be expressed as

$$\mathscr{R} = \mathscr{F}(2m\pi^2) \tag{7-38}$$

Thus a functional relationship between the finesse and the resolving power is achieved. In general, however, $\mathscr{R}$ is also proportional to the electrical separation $\eta' d$ of the two surfaces of the plate, which was contained in δ' and normalized by the mean wavelength.

The term resolving limit is generally associated with $\Delta k/2\pi$, which is the inverse of the product of the resolving power and the mean wavelength. Thus resolving limit has the dimensions of inverse length. Resolving limits on the order of 1×10^{-2} cm^{-1} are common.

Fabry–Perot Interferometer

If the angle of incidence in the multiple-beam interferometer approaches zero or normal incidence, the condition of the Fabry–Perot interferometer is achieved. In early work the instrument consisted of two glass or quartz plates, each having plane surfaces as shown in Fig. 7-11. The inner surfaces are coated with films that produce very high reflectivities.†

From the analysis in the previous section, it is clear from Eq. (7-32) that as R approaches unity, the transmission properties of this instrument are strongly dependent on the separation d, and that there will be many separations, occurring every half wavelength that allow transmission. In what follows, we shall refer to these as allowed axial modes. The sharpness of these modes is related to the finesse of the system, and the resolving limit essentially determines the axial-mode width. Numerical examples will be given at the end of this chapter.

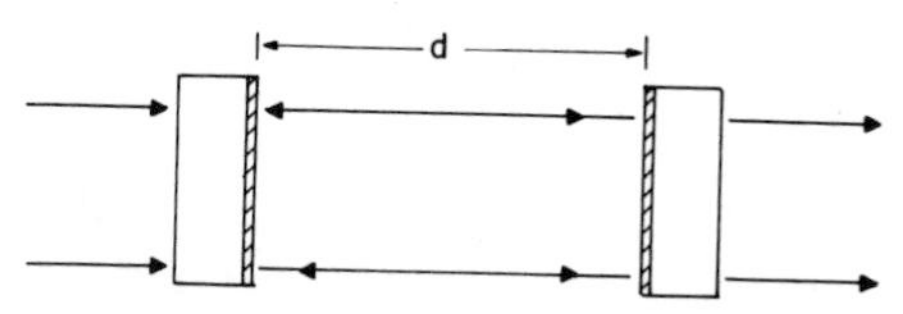

Fig. 7-11. A Fabry–Perot interferometer using plane-parallel mirrors separated a distance d.

So far the analysis of interferometers has been from a ray-optic point of view. It, however, does not give a good description of the transverse spatial variation of these allowed axial modes. To analyze the transverse spatial properties which are greatly affected by the diffraction properties of the edges‡ of the interferometer plates, a full-wave analysis has to be done.

In laser systems, the term "cavity" is used. The use of this term is potentially confusing if one does not recognize that these are open cavities. That is, the cavities are not like closed microwave cavities. The open-cavity configuration has associated diffraction losses at the edges. It will turn out in the analyses that follow that these diffraction losses will dominate the losses of the laser cavities.

From a conceptual point of view, one could think of these cavities and the peaks in transmission being associated with a resonance condition or standing-wave condition. In fact, Weinstein [58] used this notion in analytically describing many such open cavities.

† In laser cavities the reflectances can approach and in some cases exceed 99 percent.

‡ The quartz plates are not the infinite in extent plates assumed in the multiple-beam interferometer analysis. They are of finite size which in lasers is a few millimeters.

Fox and Li Analysis

One approach to the problem of analyzing a resonant system of reflecting mirrors with diffraction losses at each edge is to treat it as a succession of opaque, absorbing screens with identical holes through the system [64]. One simply allows the first hole to be uniformly excited in amplitude and phase, and then with the use of a digital computer, evaluate the Rayleigh–Sommerfeld equation for each successive aperture using the result of the field in the previous aperture to be the excitation for the next aperture. This process is continued until the amplitude and phase distribution does not change by more than a complex constant.

This was the analysis procedure used by Fox and Li. The system is shown in Fig. 7-12. The numerical analysis proceeded for q iterations, where the $q + 1$ iteration differed from q by only a complex constant. It turned out in their method that 300 iterations were required and the distributions were nearly gaussian in form. This amounted to basically finding a steady-state solution to Fredholm integral equation.

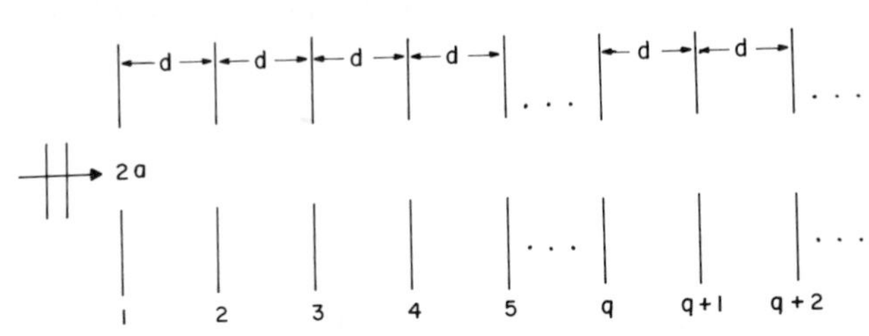

Fig. 7-12. The dual analog of the Fabry–Perot interferometer. A sequence of perfectly absorbing screens each having a hole of diameter $2a$. [Adapted with permission from *The Bell System Technical Journal*, Copyright (1961), The American Telephone and Telegraph Company.]

To describe the analysis, the Rayleigh–Sommerfeld equation, Eq. (2-1), is used. In terms of this system, it is written

$$U_2(x', y', z') = \frac{-j}{\lambda} \int_{-b}^{b} \int_{-a}^{a} U_1(x, y) \frac{e^{jkR}}{R} \, dx \, dy \tag{7-39}$$

If now the functional form of the field on the $q + 1$ aperture is assumed to differ by only a complex constant σ from the functional form of the field in the q aperture, then Eq. (7-39) is written as

$$U_{(q+1)}(x', y', z') = -j \frac{\sigma}{\lambda} \int_{-b}^{b} \int_{-a}^{a} U_q(x, y) \frac{e^{jkR}}{R} \, dx \, dy \tag{7-40}$$

where $U(\ )$ is a functional form that does not change in transverse variation from pass to pass.

Equation (7-40) is now in the form of a known integral equation for which a normal set of eigenfunctions can be found. Thus these eigenfunctions, which constitute a solution, are also the normal-mode functions for the allowed modes. The complex constant σ is proportional to the product of the eigenvalues for each transverse direction in the system.

A physical meaning can be associated with the eigenvalue. Since it represents the change in the steady-state solution from aperture to aperture, the real part of σ can then be associated with the diffraction loss† around the edges of the mirrors. Fox and Li found by their numerical procedure that the loss per transit followed curves like those shown in Fig. 7-13, where different mirror geometries are presented. Loss is plotted versus Fresnel number $a^2/l\lambda$ and shows that for Fresnel numbers on the order of 100, the loss is well below 1 percent and less than the absorption losses associated with the mirrors, which may be only a fraction of 1 percent.

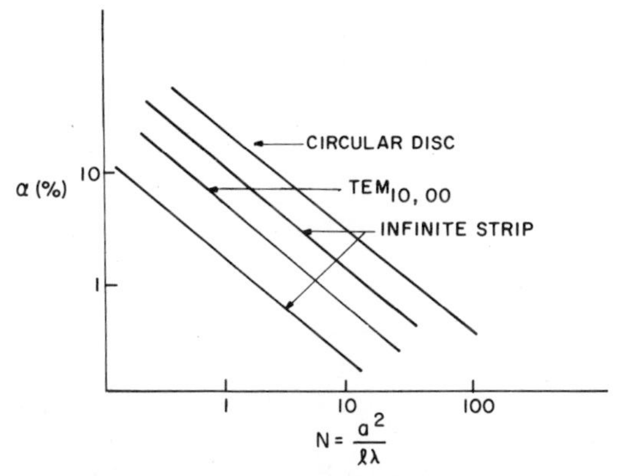

Fig. 7-13. Power loss per transit versus the Fresnel number N. [Adapted with permission from *The Bell System Technical Journal*, Copyright (1961), The American Telephone and Telegraph Company.]

Analytical Solution, Boyd–Gordon Approach

The Fox and Li approach was numerical and lacks the flexibility and usefulness of an analytic representation. It is possible to solve Eq. (7-40) analytically as shown by Boyd and Gordon [62]. They analyzed a confocal

† This assumes that the loss is mostly dominated by the diffraction loss.

multimode (transverse direction) resonator formed by two spherical reflectors spaced by their common radius of curvature. In an analytical way, the analytic procedure was to develop the eigenfunctions for the natural modes of the geometry. The mode patterns and diffraction losses are also obtained by this method. The confocal spacing was found to be an optimum in the sense of minimum diffraction losses and mode volume.

The formulation of the problem is essentially the same as Fox and Li except that now a product solution for $U(x, y)$ is assumed. Thus represent $U(x, y)$ as $E_0 f_m(x) g_n(y)$, which modifies Eq. (7-40) to

$$E_y(x', y', z') = \frac{E_0}{j\lambda} \iint_A f_m(x) g_n(y) \frac{e^{ikR}}{R} \, dx \, dy \tag{7-41}$$

where R is the distance between P and P', and θ is the angle between PP' and the normal to the aperture spherical surface as shown in Fig. 7-14.

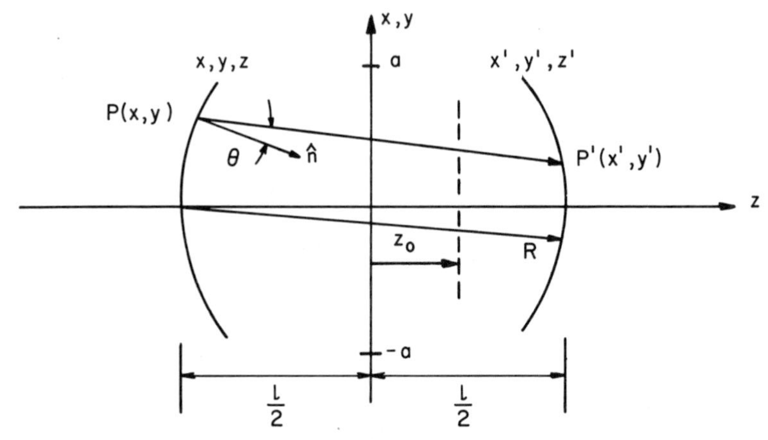

Fig. 7-14. Confocal resonator geometry using spherical reflectors. [Adapted with permission from *The Bell System Technical Journal*, Copyright (1961), The American Telephone and Telegraph Company.]

The normal modes or eigenfunctions of the confocal resonator are obtained by requiring that the field distribution over (x', y') be functionally the same, to within a constant, as the field distribution over the (x, y) aperture. Thus $E_y(x', y')$ is written as

$$E_y(x', y') = E_1 f_m(x') g_n(y') \tag{7-42}$$

where $E_1 = \sigma_m \sigma_n E_0$ and σ_m, σ_n are complex constants, representing both the amplitude and the phase changes. Thus

$$\sigma_m \sigma_n f_m(x') g_n(y') = -\frac{j}{\lambda} \int_{-b}^{b} \int_{-a}^{a} \frac{e^{ikR}}{R} f_m(x) g_n(y) \, dx \, dy \tag{7-43}$$

where the electric field in the xz plane is assumed to be zero.

For small apertures, the Fresnel approximation is used† giving an approximate form for R as

$$\frac{R}{l} \simeq 1 - \frac{xx' + yy'}{l^2} + \frac{x^2 + y^2}{2l^2} + \frac{x'^2 + y'^2}{2l^2} + \cdots \tag{7-44}$$

If $a^2/l\lambda \gg a^2/l^2$, then the third and fourth terms can be neglected, and after a change of variables,

$$C \equiv a^2k/l = 2\pi(a^2/l\lambda), \qquad X = x\sqrt{C}/a, \qquad Y = y\sqrt{C}/a \tag{7-45}$$

and substitution of $F_m(X) = f_m(x)$, and $G_n(Y) = g_n(y)$, the following normal equation is obtained:

$$\sigma_m \sigma_n F_m(X) G_n(Y) = \frac{ie^{ikl}}{2\pi} \int_{-\sqrt{C}}^{\sqrt{C}} F_m(X') e^{iXX'} \, dX' \int_{-\sqrt{C}}^{\sqrt{C}} G_n(Y') e^{iYY'} \, dY' \tag{7-46}$$

This integral equation can be studied by noting that it is the product of two integral equations which are of the homogeneous Fredholm type of the second kind. Since the kernel is $e^{iXX'}$, they are also of the form of the finite Fourier transform. Slepian and Pollack studied the following integral equation

$$F_m(X) = \frac{1}{\sqrt{2\pi}\, X_m} \int_{-\sqrt{C}}^{\sqrt{C}} F_m(X') e^{iXX'} \, dX' \tag{7-47}$$

where the resulting eigenfunctions were wave functions in the prolate spheroidal coordinates [66]. In general, the form is like

$$F_m(C, N) \propto S_{0m}(C, N) \tag{7-48}$$

and

$$X_m = (2C/\pi)^{1/2} i^m R_{0m}^{(1)}(C, 1) \qquad m = 0, 1, 2, \ldots \tag{7-49}$$

where $S_{0m}(C, N)$ and $R_{0m}^{(1)}(C, 1)$ are the angular and radial wavefunctions, respectively, and

$$N = X/\sqrt{C} = x/a \qquad N = y/a \qquad \text{for} \quad G_n(Y) \tag{7-50}$$

† See Chapter II for further discussions on the Fresnel approximations.

The important points of this analytical solution are that

(a) the eigenfunctions are real; therefore, the reflecting surface is of constant phase;

(b) the eigenvalues are $\sigma_m \sigma_n = X_m X_n i^{-ikl}$, where the phase shift per pass is determined by the phase of the eigenvalue.

Thus the resonance round-trip phase shift is

$$2\pi q = 2 \mid (\pi/2) - kl + (m + n)(\pi/2) \mid \tag{7-51}$$

from which a resonance condition can be written

$$4l/\lambda = 2q + (1 + m + n) \tag{7-52}$$

Note that considerable degeneracy exists, implying that many frequencies can oscillate simultaneously.

From Eq. (7-52) and the fact that a multiple of 2π phase per round trip must exist in the cavity, the separation of these degenerate modes can be calculated by evaluating the frequency difference between the $(q + 1)$th and qth modes. Using $\lambda = c/f$, we see that this becomes

$$(q + 1)\pi - q\pi = (2d\pi/c)(f_{q+1} - f_q) = (2\pi d/c)\, \Delta f \tag{7-53}$$

Canceling π from both sides of the equation, Δf becomes

$$\Delta f = c/2d \tag{7-54}$$

An example can be used to provide insight on the kinds of numbers involved. Thus for d equal to 1 m, Δf is[†]

$$\Delta f = 150 \text{ MHz} \tag{7-55}$$

In general, for longer and longer cavities the frequency spacing becomes smaller and smaller. For typical cavities in the He–Ne configuration, d has to approach 10 cm before a single mode will operate in the He–Ne doppler broadened line width of about 1500 MHz.

Since the axial field was assumed to be negligible, the modes can be designated by a notation similar to the transverse electromagnetic mode of microwave cavities, except that an axial-mode number is added. Thus the designation becomes TEM_{mnq}, where $m, n = 0, 1, 2, \ldots$. The first two indices m and n refer to x and y variation. The third indice q refers to the z axis or axial-mode variation. In general, the form of a few of the

[†] In this case the velocity of light c is 3×10^8 m/sec.

eigenfunctions can be represented by simple gaussian–hermite functions, such as,

$$F_0(C, N) = e^{(-1/2)CN^2} \tag{7-56a}$$

$$F_1(C, N) = \sqrt{\pi C}\, N e^{(-1/2)CN^2} \tag{7-56b}$$

$$F_2(C, N) = (2CN^2 - 1)e^{(-1/2)CN^2} \tag{7-56c}$$

The functions described in Eq. (7-56) can be plotted to show how typical mode patterns would appear. Basically, only the transverse mode structure can be illustrated, and these are shown in Fig. 7-15. It is important to gain an idea of this structure so that an intuition is developed for recognizing modal structure as distinct from other forms of spatial modulation, such as would be produced by a grating.

The exponential dependence on CN^2 leads to the definition of spot size W_s, which is independent of the mirror radius a. It can be shown that [65]

$$W_s = (l\lambda/\pi)^{1/2} \tag{7-57}$$

The spot size is the point at which the field amplitude falls to e^{-1} of its value along the axis of symmetry. Thus the only effect of increasing the reflector radius a is to reduce the diffraction losses.

One final point of emphasis is that the prolate spheroidal functions have the unique property that they are orthogonal over two ranges. One is finite $(-1, 1)$, and the other is infinite $(-\infty, \infty)$. This property is very useful when dealing with bandlimited systems of any type and simplifies the theoretical analysis of general finite aperture systems.

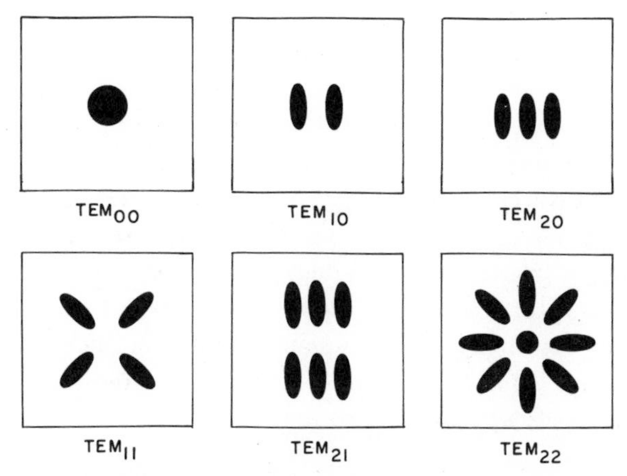

Fig. 7-15. Typical conceptual modal patterns obtained from a laser using a Fabry–Perot interferometer.

Stability of Modes

Once the existence of modes in these open-resonator geometries is known, it follows quite naturally to question which mirror spacings and curvatures will support such modes. Boyd and Kogelnik [67] studied this problem, along with others, to derive analytical expressions that represent the criteria for mode stability.

In this case, the word "stability" has a restricted use in that it refers to whether there are spacings and curvatures that will naturally support resonances in the open-resonator geometries. This is quite a crucial question since these systems must operate in low-loss regions and the sensitivity of these regions to slight changes in the radius of curvature is very important. Fabrication problems and variations in the mirror support mechanisms may cause these systems to detune and essentially be attenuators rather than resonators.

The question can be examined by writing a simple expression for the amplitude of the lowest-order mode in the region between the mirrors,

$$E = \frac{AW_0}{W(z)} \exp\left[-\frac{r^2}{W(z)^2}\right] \tag{7-58}$$

where

$$W(z) = W_0[1 + (z/z_0)^2]^{1/2} \tag{7-59}$$

and

$$z_0 = \pi W_0^2/\lambda \tag{7-60}$$

In this case, $W(z)$ represents a description of the beam radius. It is a function of axial position and tends to diverge with increasing z. The factor z_0 is a parameter that is related to the Fresnel number of the minimum beam size or "spot size." It essentially represents the region where the beam begins to appear as a diverging spherical wave. It is not necessarily related to the location of the mirrors, but it can be for some geometries.

If one plots $W(z)$ versus z for values of z between $\pm d$, it is found that the surfaces are nearly spherical. Thus spherical mirrors can be placed at appropriate points to support a resonance condition. Such a configuration is shown in Fig. 7-16. The radius of curvature of these surfaces can be calculated from the approximate formula

$$R(z) \cong \frac{W(z)}{dw/dz} \tag{7-61}$$

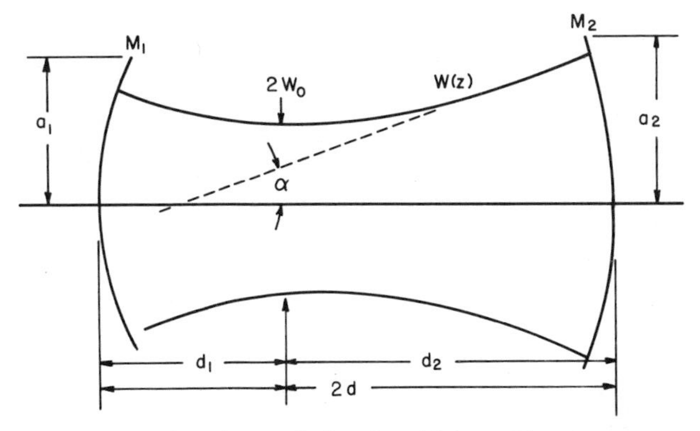

Fig. 7-16. Geometry for the analysis of stable/unstable resonator configurations. [Adapted with permission from *The Bell System Technical Journal*, Copyright (1962), The American Telephone and Telegraph Company.]

Differentiating Eq. (7-59) and substituting into Eq. (7-61), we obtain R for $z = d/2$:

$$R = \frac{d}{2}\left[1 + \left(\frac{2z_0}{d}\right)^2\right] \tag{7-62}$$

Using this expression, the radius of curvature can be written for each mirror:

$$R_1 = \frac{d_1}{2}\left[1 + \left(\frac{2z_0}{d_1}\right)^2\right] \tag{7-63a}$$

$$R_2 = \frac{d_2}{2}\left[1 + \left(\frac{2z_0}{d_2}\right)^2\right] \tag{7-63b}$$

Equations (7-63) are used to solve for d_1 and d_2, and then expressions for the appropriate spacing of each mirror are written. These expressions are

$$d_1 = R_1 \pm (R_1^2 - 4z_0^2)^{1/2} \tag{7-64a}$$

$$d_2 = R_2 \pm (R_2^2 - 4z_0^2)^{1/2} \tag{7-64b}$$

and

$$2d = d_1 + d_2 \tag{7-65}$$

If Eqs. (7-65) and (7-64) are combined, a solution for z_0^2 in terms of the generally given parameters R_1, R_2 and d can be written

$$z_0^2 = \frac{d(R_1 - d)(R_2 - d)(R_1 + R_2 - d)}{(R_1 + R_2 - 2d)^2} \tag{7-66}$$

An important special case of this result is the confocal configuration defined by $R_1 = R_2 = d$, leaving

$$z_0 = d/2 \tag{7-67}$$

Using Eq. (7-60), we can write the "spot size" for this configuration as

$$W_0 = (d\lambda/2\pi)^{1/2} \tag{7-68}$$

which leads to a beam diameter of

$$W(d/2) = \sqrt{2}\, W_0 \tag{7-69}$$

As an illustration, the spot size for a spacing of 1 m, and $\lambda = 632.8$ nm can be evaluated, giving

$$W_0 \cong 0.32 \text{ mm} \tag{7-70}$$

and

$$W \cong 0.45 \text{ mm} \tag{7-71}$$

In general, the position of the minimum waist size can be calculated by substituting the determined value of z_0 into Eq. (7-64) and then solving for d_1 and d_2. In general, this minimum beam size is not in the center of the cavity. Only for the confocal case is it in the center.

Stability Conditions

Using Eqs. (7-64) and (7-65), we can derive the expression for the values of d that correspond to a resonance:

$$2d = d_1 + d_2 = R_1 + R_2 \pm (R_1^2 - 4z_0^2)^{1/2} \pm (R_2^2 - 4z_0^2)^{1/2} \tag{7-72}$$

To obtain the various allowed resonator systems, the parameter z_0 is varied. The range of this variation is, however, restricted. The basis of this restriction is that only real values of d have any physical meaning. It is fairly easy to see that the case that $R_2 > R_1$ and d real, $2z_0$ is restricted to the range

$$0 \leq 2z_0 \leq R_1 \tag{7-73}$$

For this range, d is restricted to two ranges,

$$0 \leq d \leq R_1 \tag{7-74a}$$

and

$$R_2 \leq d \leq R_1 + R_2 \tag{7-74b}$$

Using Eq. (7-74b), we can derive a more general bounding condition on the stable and unstable region. If $d + R_1R_2/d$ is subtracted from both sides of the right-hand inequality and the inequality is reversed with minus one, the following is obtained:

$$R_1R_2/d > d - (R_1 - R_2) + R_1R_2/d \tag{7-75}$$

Dividing through both sides with d/R_1R_2 leaves

$$d^2/R_1R_2 - d/R_1 - d/R_2 + 1 < 1 \tag{7-76}$$

which when factored, reduces to

$$0 \leq (d/R_1 - 1)(d/R_2 - 1) \leq 1 \tag{7-77}$$

This is the general equation for a hyperbola. The bounding hyperbolas are plotted as heavy lines in Fig. 7-17, and the dashed region is the unallowed or unstable region. For purposes of illustration, several common resonator configurations can be listed along with the rational concerning its use:

Confocal $d = R_1 = R_2$

This case is undesirable because it can easily move into a region of instability.

Parallel plane $R_1 = R_2 = \infty$

This case is also marginally stable as it is at the border of the bottom unstable region.

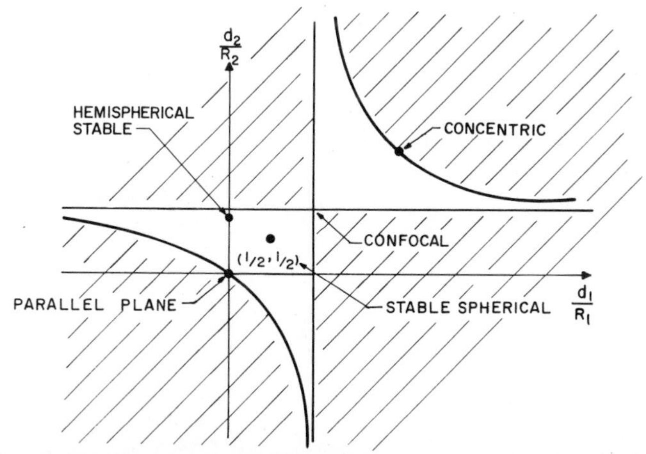

Fig. 7-17. Stability diagram for various reflector spacings. The cross-hatched area represents unstable or high-loss configurations. [Adapted with permission from *The Bell System Technical Journal*, Copyright (1962), The American Telephone and Telegraph Company.]

Concentric resonator $R_1 = R_2 = d/2$

This case is the dual of the parallel plane and is also undesirable.

Stable spherical $d_1/2 < R_1 = R_2 < d_2$ and $d_1 < R_1 = R_2 < \infty$

This case is stable since it lies within the stable region. The case $R = R = 2d$ is shown in the figure.

Hemispherical-stable $R_1 = \infty$, $d_2 < R_2$

This is a popular configuration for stable modes and is often used for its unwanted mode suppression capability.

These cases are shown in Fig. 7-17.

Diffraction Losses

In general, the diffraction losses are determined from the real part of the eigenvalues and written as

$$\alpha_D = 1 - |\sigma_m \sigma_n|^2 = 1 - |X_m X_n|^2 \tag{7-78}$$

Boyd and Gordon [65] found an approximate expression for the losses for various values of the Fresnel number N:

$$\alpha \simeq \pi^2 2^4 N e^{-4\pi N}, \qquad \text{large } N \tag{7-79a}$$

$$\alpha \simeq 1 - \pi^2 N^2, \qquad \text{small } N \tag{7-79b}$$

These formulas can be used to approximate typical losses. For a typical mirror radius of 7 mm, N is about 250, which is a case of large N, leading to a value of α which is to good approximation zero. Thus with diffraction losses of this order, these open-resonator systems will oscillate with very little gain in the medium. Typically, He–Ne lasers have a gain of 0.005 or one-half of 1 percent per pass.

Cavity Q

A related concept that enables a better understanding of losses is the idea of cavity Q [1]. The Q of a cavity is the energy stored per cycle over the average energy lost

$$Q = \omega_0 U / W_L \tag{7-80}$$

where the energy stored U is related to the energy density U_v and volume by

$$U = \pi W^2 d U_v \tag{7-81}$$

The losses are primarily due to diffraction and absorption losses in the mirrors. Thus the power lost is found by calculating the power transfer from the group velocity c and the energy density U_v. The fraction of transferred power lost is designated by ξ, leaving

$$W_L = \pi W^2 c U_v \xi \tag{7-82}$$

Thus Q becomes

$$Q = \frac{\pi W^2 d U_v \omega_0}{\pi W^2 c U_v \xi} = \frac{2\pi d}{\lambda \xi} \tag{7-83}$$

To illustrate, consider the effect of $\frac{1}{2}$ percent mirror loss and a 50-cm cavity length at a wavelength of 6328 Å,

$$Q = \frac{(2\pi)(0.5)}{(0.005)(0.6328 \times 10^{-6})} \cong 1 \times 10^9 \tag{7-84}$$

This Q is very large and is generally at the limit of capability with state-of-the-art mirrors. It is, also, much large than the Q that can be obtained in the microwave region, where a Q of 1000 to 10,000 is very good.

Problems

1. Derive Eq. (7-18).

2. Derive Eqs. (7-23), (7-24), (7-31), and (7-32).

3. Derive the finesse $\mathscr{F}$, the resolving power $\mathscr{R}$, and the resolving limit for reflectance $R = 0.995$, and wavelength $\lambda = 6328$ Å. Assume an air-spaced, lossless, Fabry–Perot interferometer.

4. For $d = 2$ m, $R_1 = 3$ m and $R_2 = 4$ m, calculate d_1, d_2, z_0, W_0, and $W(d_1)$ and $W(d_2)$.

5. Derive Eqs. (7-64).

6. Using the definition of curvature established in calculus, show the limits of applicability of Eq. (7-61).

7. Derive Eq. (7-62) using (7-61) and (7-59).

8. Plot $(W_{\text{min}}/W_{0\ \text{confocal}})$ for various values of d/R; $0 \leq d/R \leq 2$.

9. Find the axial frequency spacing for a cavity that is 30 cm long and mirrors of 15 mm diameter, $\lambda_0 = 6328$ Å.

10. Find the Q of the cavity described in Problem 9 if $\xi = 0.4$ percent and $\lambda_0 = 6328$ Å.

11. Determine whether the following cavities are stable or unstable.

a. $d_1 = R_1,\ d_2 = R_2$
b. $d_1 = \frac{3}{2}R_1,\ d_2 = \frac{3}{2}R_2$
c. $d_1 = -R_1/2,\quad d_2 = R_2/2$
d. $d_1 = R_1/2,\ d_2 = -R_2/2$
e. $d_1 = R_1/2,\ d_2 = 3R_2/2$

12. Draw mode patterns for the following modes: TEM_{32}, TEM_{30}, TEM_{41}.

13. Evaluate the Q for the cavity of Problem 3, using the approximate result for Q of $f_0/\Delta f$.

Chapter VIII

HOLOGRAPHY

Introduction

The desire to reconstruct the entire image of a recorded object has long been a quest in optics. Dennis Gabor in 1948 described a new reconstruction method for improving the resolution limit in electron microscopy. Although he could not demonstrate this technique with electron waves, he demonstrated the idea with visible light [68–70]. His demonstration, however, allowed all the reconstructed images and distortion terms to appear superposed. The corrupted images, which appeared in his demonstration, only served to detract from the acceptance of the method. It was not until the early 1960s that Leith and Upatnieks [71–73] were able to demonstrate a method for separating the images and distortion terms. This addition to the basic reconstruction method then provided a springboard for acceptance and use of the method [64–67].

The term "holography" as applied to wave-front reconstruction arises from the synthesis of two Greek words—"holos" and "gram." The word "holos" means *entire*, and the word "gram" is a derivative of the Greek words, gramma, or graphein, meaning *vivid writing*, or that which is written. The explicit syntax for the word "holography" is then "entire picture."

The fundamental addition to the process of imaging that comes in this method is the addition of the phase information about the object. Prior to holography, pictures were simply a brightness rendition, yielding only information about the object amplitude distribution. The successful addition of phase to the viewed image made total reconstruction of the object wave possible. With phase information such characteristics as depth and parallax are presented and give the viewer the natural perception of three dimensions.

It would seem that uncorrupted three-dimensional imaging involved the discovery of something entirely new. This was not quite true, as will be shown later, since most of the technique had already been used in the development of the matched filter [39]. The essential difference is whether a reference wave or object wave is used in reconstruction. In Chapter V it was shown how an interferogram, illuminated with the object wave at an appropriate angle, produced an image of a spot to indicate the presence of the original signal. The converse of this process was recognized by Leith and Upatnieks [71–73] as a method for reproducing or reconstructing the object wave if a reference wave illuminated the interferogram at an appropriate angle.

To this point the development, briefly described, provides the framework for much of the novelty and popularity associated with holography. A second aspect of holography, which does not focus on the three-dimensional viewing, is the use of the entire reconstructed image as a complex wave form in an optical data processor. This application [71, 78–86] has led to such devices as holographic lenses [86], lens corrector elements [87], simplified optical Fourier processors [82], and other transform devices [88]. Thus a hologram can be treated as an element in an optical processor. This point of view will be emphasized in much of what follows.

Generation of Phase Information

Part of the understanding of what a hologram is depends on the understanding of how the phase information is added to the reconstructed image. The phase and amplitude information is derived from the scattered waves that arise when a reference wave propagates through a grating structure that contains both the original scene information and the reference wave information. In this sense it can be viewed as a modulation process where the reference wave is spatially modulated by an interferogram. A process, which may be more familiar, is that of the heterodyne detection in radio. In heterodyne detection a local oscillator is used to demodulate the incoming rf and present the information which is encoded on the envelop of the rf wave.

The analog of this process is possible in optics because film is the basic element. The dependence of film exposure on the total intensity† implies that the density of silver will depend on the absolute value or modulo

† Intensity can be used if the intensity is constant over the period of exposure.

squared of all waves incident on the holographic plate during the construction process. The mathematical model of such a process contains summed terms that are complex conjugates of each other. This arises because physical constraints through the intensity dependence require that the representation be totally real. This constraint is satisfied, since a real number can be obtained from a set of complex numbers if sums occur over complex conjugate pairs, like

$$2\,\mathrm{Re}\{A\} = A + A^* \tag{8-1}$$

Herein lies the reason why a complex field (phase and amplitude) can be obtained in the output. If a real number like $\mathrm{Re}\{A\}$ is multiplied by another complex number B, then a pair of complex numbers $BA + BA^*$ is obtained. Since A and A^* will modify B differently, the output of this kind of modulation process is contained in two terms or two waves, which may appear at two different spatial locations. These two terms always appear in a process that uses only "real" elements. Physically B can be thought of as an incident plane wave. It is complex, meaning it has amplitude and phase. The plane wave is modulated by the interferogram, or rather the plane wave is scattered by the interferogram much as a wave is scattered by a simple grating. This scattering process gives rise to conjugate pairs of waves whose phase and amplitude have been shaped by the interferogram. It is convenient at this point to think of the interferogram as just a collection of gratings. The choice of what gratings are in the collection and what depth of modulation applies is determined by the original input image.

The fact that the conjugate pairs of waves appear at different spatial locations is due to the grating effect.† The scattering or diffraction of the reference wave by the composite of gratings splits the reference wave into two off-axis and two on-axis spatial components. This then provides access to the complex wave representing the original input scene and its conjugate. The reconstructed object wave is called a virtual image, and the conjugate wave is a real image.

It is rather interesting that Gabor [68, 69, 89] did not recognize in his original work this method of spatially separating the pairs of reconstructed images. He recognized their pair-wise nature, but in his original setup they were superimposed as shown in Fig. 8-1. It remained for Leith and Upatnieks [71–73] to recognize that nonnormal incidence could be used to separate the components as shown in Fig. 8-2, the conjugate pair of reconstructed images at symmetrical angles to the reference wave vector. The

† See Chapters II and V for problems and explanations of grating systems.

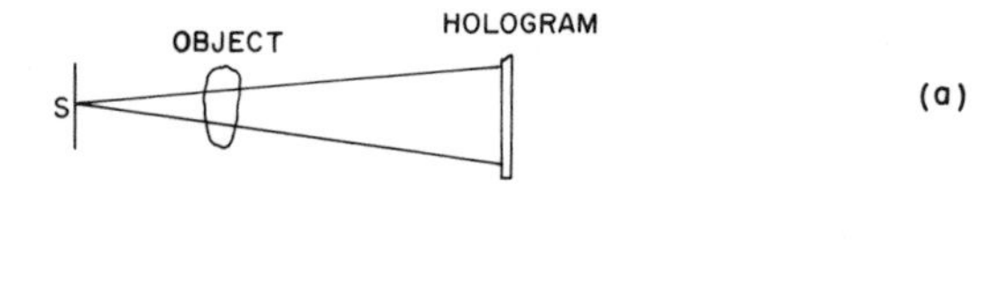

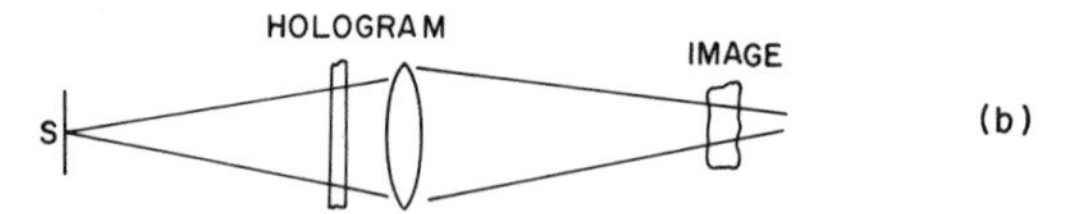

Fig. 8-1. Gabor hologram, semitransparent object. (a) Formation. (b) Reconstruction.

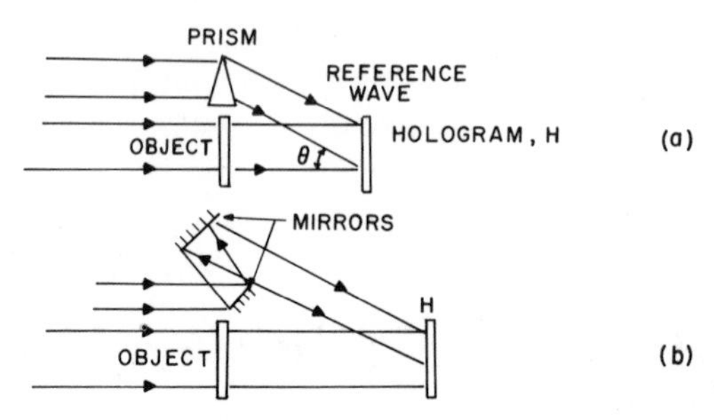

Fig. 8-2. Basic Leith and Upatnieks hologram—sideband hologram. (a) Prism method. (b) Mirror method.

virtual image appears in focus, and the real image generally is defocused. It will be shown later that illumination with a conjugated reference wave will provide an infocus real image. First, however, details of the interferogram construction will be presented.

Formation of the Interferogram

The interferogram is a density profile that corresponds to the intensity distribution at the hologram plane in the original arrangement. In Chapter V the interferogram development for matched filters was presented in terms of the exponential representation. Thus Eq. (5-9) is repeated to illustrate the model for the interferogram.

$$I = |U_1 + U_2|^2 = A_1^2 + A_2^2 + A_1 A_2^* \exp[i(\psi_1 - \psi_2)] + A_1^* A_2 \times \exp[-i(\psi_1 - \psi_2)] \tag{8-2a}$$

$$U_1 = A_1 e^{i\psi_1} \tag{8-2b}$$

$$U_2 = A_2 e^{i\psi_2} \tag{8-2c}$$

This equation represents the intensity of the two signals at the hologram plane, where U_1 is the information signal and U_2 is the reference wave. The exposure determined by the intensity distribution of Eq. (8-2a) can result in an appropriate amplitude transmission coefficient if an appropriate gamma is chosen, or if a large-amplitude reference wave is used. Both of these cases will be discussed in what follows. Thus the exposed piece of film is developed, and an amplitude transmission coefficient results:

$$t_a(x, y) = CI^{-\gamma/2} \tag{8-3}$$

Substituting Eq. (8-2a) and using Euler's relations, we obtain the following form for $t_a(x, y)$:

$$t_a(x, y) = C[A_1^2 + A_2^2 + 2A_1A_2 \cos(\psi_1 - \psi_2)]^{-\gamma/2} \tag{8-4}$$

Were it possible to develop with a gamma of minus two and then illuminate with a reference wave corresponding to U_2 as shown in Fig. 8-3, then the following wave U_3 would be emitted from the reconstruction process

$$U_3(x, y) = U_2t_a = A_2^2U_1(x, y) + (A_1^2 + A_2^2)A_2e^{i\psi_2} + A_1A_2^2 \times \exp[-i(\psi_1 - 2\psi_2)] \tag{8-5}$$

The first term, multiplied by the constant A_2^2, is our desired total reconstructed image. It contains both the amplitude and the phase of the original object wave. If separated from the other images present, it would appear to the viewer as the original image in all three dimensions. The perception of the three dimensions is a result of the fact that various points on the object have a phase relationship with respect to each other. The apparent depth perception and parallax effects are established by these relative phase differences.

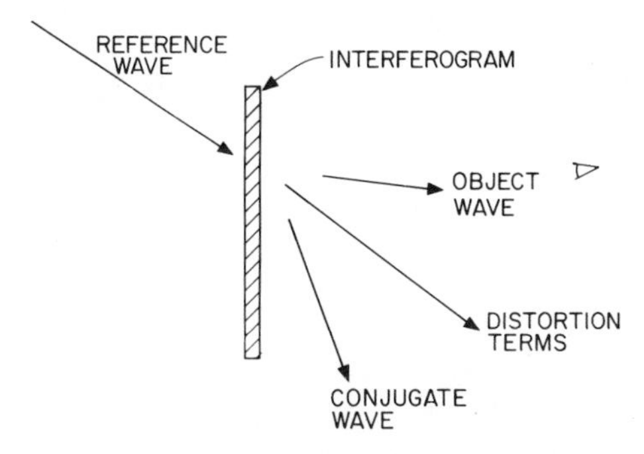

Fig. 8-3. Reconstruction waves for sideband holography.

The second and third terms in parentheses in Eq. (8-5) correspond to waves that are in the direction of the reference wave and are generally referred to as distortion terms. If the reference wave amplitude is large enough, the second term is negligible compared to the third term and thus dropped. The fourth term in Eq. (8-5) is like the first term in that it is proportional in amplitude to the desired image, but it has the conjugate phase. It also scatters in a direction that is symmetrical but opposite to the reference wave axis. This conjugate wave gives rise to the so-called twin or real image.

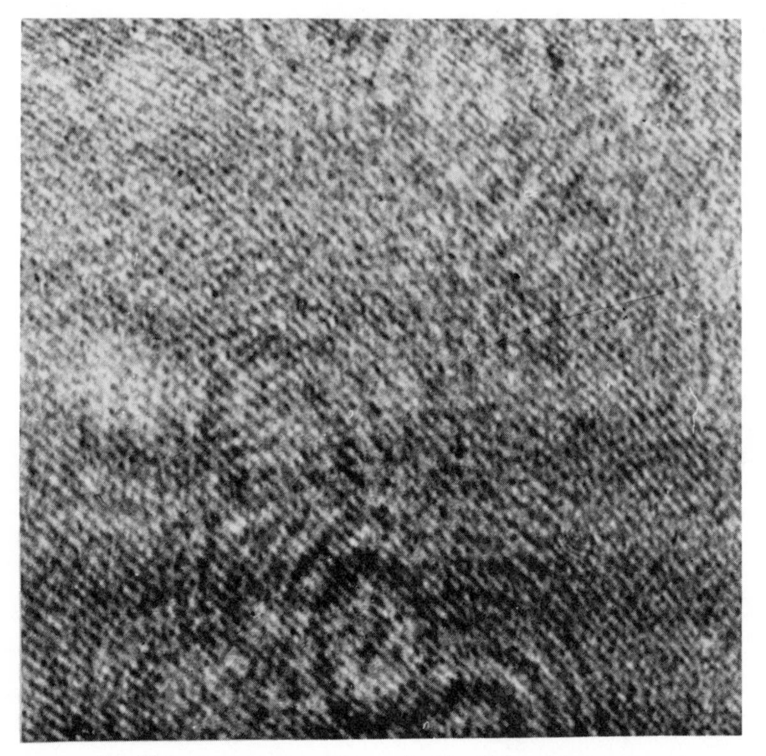

Fig. 8-4. Interferogram for a simple sideband hologram.

Figure 8-4 shows a sample hologram or interferogram. Note the grating characteristic that seems to exist throughout the plane. This is the sinusoidally varying density pattern that gives rise to the scattered images. The ring patterns are actually coherent noise. The coherent noise is caused by dust particles that appear on lenses and other interfaces.

Arbitrary Gammas

The achievement of gamma equal to minus two is not possible in the single-step hologram construction process.† In general, the nonlinear process is linearized by using a large reference wave amplitude. The large reference wave amplitude permits the use of the binomial expansion approximation. Thus consider a modification of Eq. (8-4) for the case that $A_2 \gg A_1$. Factoring out A_2^2 and rearranging give

$$U_3(x, y) = A_2^{1-\gamma} e^{i\varphi_2} \left[\left(1 + \frac{A_1^2}{A_2^2}\right) + 2\frac{A_1}{A_2} \cos(\varphi_1 - \varphi_2) \right]^{-\gamma/2} \tag{8-6}$$

Use of the binomial expansion and neglect of the term A_1^2/A_2^2 reduces Eq. (8-6) to

$$U_3(x, y) \cong A_2^{1-\gamma} e^{i\varphi_2} [1 - (A_1/A_2)\gamma \cos(\varphi_1 - \varphi_2)] \tag{8-7}$$

Appropriate multiplication, expansion, and collection of terms give

$$U_3(x, y) \cong \frac{1}{A_2^{\gamma}} \left[U_2(x, y) - \frac{\gamma}{2} U_1(x, y) - \frac{\gamma}{2} U_1^*(x, y) e^{i2\varphi_2} \right] \tag{8-8}$$

The transmitted wave $U_3(x, y)$ is like the expression in Eq. (8-5), except that the amplitudes of the reconstructed images are reduced by the term $-\gamma/2A_2^{\gamma}$. This fact is not so bothersome since the contrast ratios are not affected, and reconstruction amplitudes can be increased overall. The first term is just the direct, though attenuated, reference wave. The second and third terms are the two desired virtual and real images that are simply changed by the constant multiplying factor noted above. Thus the use of a large-amplitude reference wave results in a reconstruction system that is very little different from the mathematically idealized case of gamma equal to minus two.

Reconstruction of the Object Wave

The second term in Eq. (8-8) is the desired information or object wave. The out-of-focus object wave, the real image, appears at another angular position with respect to the reference wave axis, and its position is described

† See Chapter VI for a description of how minus gammas are obtained.

mathematically by the multiplication phase term $\exp[i2\psi_2]$. The virtual image is a reconstruction of the object wave to the extent that the amplitude and phase recorded over the field of view of the holographic plate represent the object. In addition, the form of Eq. (8-7) shows that the virtual image reconstructed wave is one part of a spatial modulation process that presents the object phase in relationship to the reference wave phase.

The referral to reference wave takes on special meaning when it is noted that ψ_2 equals $\mathbf{k}_2 \cdot \mathbf{r}$ or $kr \cos \theta_2$, where θ_2 is the angle between direction of the reference wave and the coordinate system. Thus $(\psi_1 - \psi_2)$ is the difference of two cosine terms which is related to the product of sine functions whose arguments are sum and differences of the arguments of each cosine function. Thus the reconstructed waves propagate in directions relative to the direction of the reference wave vector.

The object wave reconstruction can also be considered an imaging process much as imaging was described in Chapters III and IV. This point of view is also consistent with how the viewing process appears. When observing the virtual image, it appears that the reconstructed object wave is emanating from the same physical points in space as where the object actually was. Thus in this sense, the portion of U_3 that represents the object wave is just a duplication of the original object wave.

Fresnel and Fraunhofer Holograms

Further delineation of the concepts of holography requires more specifics about some of the terminology. The most common phrases used are Fresnel and Fraunhofer holograms. These terms refer primarily to the distance from the object or image to the hologram plane. In Chapter II regions from an aperture were defined by the aperture size a, the wavelength λ, and distance to the observation point. The parameter of concern was the square of the aperture size divided by the wavelength λ. If the distance z_0 is very much larger than a^2/λ, then this is the Fraunhofer zone, and any hologram made this way will be referred to as a Fraunhofer hologram. If z_0 is in the near zone or Fresnel zone of the object ($z_0 \leq a^2/\lambda$), then a Fresnel hologram is obtained. The basic issue here is the diffraction properties of the input scene. This plays an important part in the quality of the reconstructed image. In particular, the Fraunhofer hologram suffers from the fact that the conjugate image interferes with the focused image. This arises because the conjugate image is located essentially at infinity and appears in the observed virtual image.

Wave-Front and Amplitude Holograms

This terminology primarily refers to the method of obtaining the reference beam. This is illustrated in Fig. 8-2. Figure 8-5 shows a method for dividing the wave front and then mixing it through a prism onto the hologram plate. This process requires that the phase across the wave front have good correlation properties. This distinction is also like the one made in interferometry, where interferometers were divided into two broad types—amplitude and wave-front interferometers.

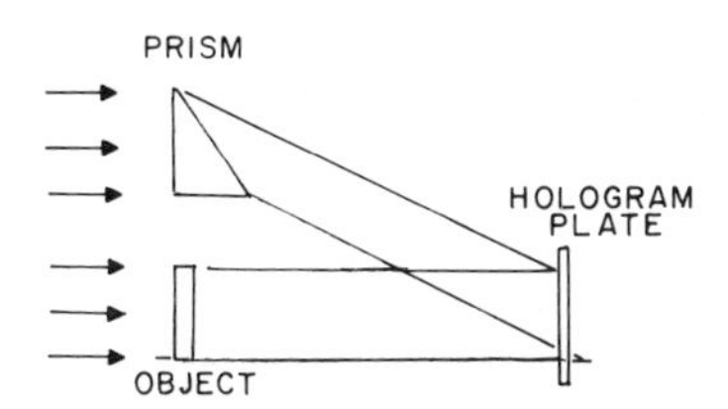

Fig. 8-5. Wave-front division for sideband holography.

Figure 8-6 illustrates a procedure for constructing an amplitude hologram. The distinction is that the amplitude across a wave front is divided into two parts, one forming the reference beam and the other the illumination for the object. In this case there is more of a requirement on the monochromatic or temporal properties of the source, since in general the mixing corresponds to two different time samples of the source waves.

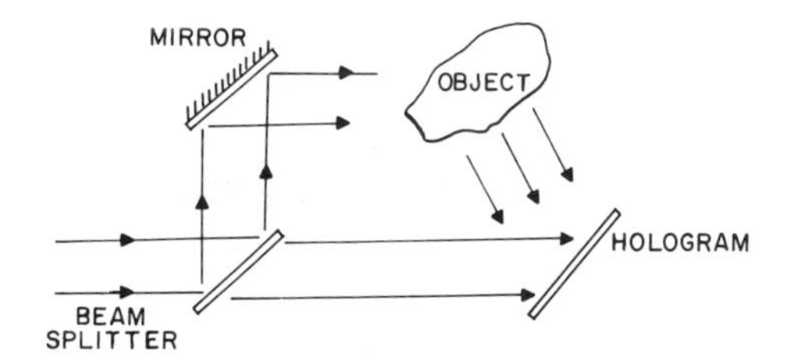

Fig. 8-6. Amplitude division for sideband holography.

Fourier-Transform Holograms

The next distinction is the type of hologram that is formed with the Fourier transform of an amplitude distribution and a point-source reference. The point-source reference partly serves the purpose of canceling some of the quadratic phase error that exists in the hologram plane. This is particularly true in the case where the Fraunhofer pattern of an object, which is proportional to the Fourier transform of the amplitude distribution, is mixed with a point-source reference.

It is also easy to see why the terms "Fraunhofer and Fourier holograms" are so easily mixed up. The Fraunhofer hologram utilizes a plane-wave reference source with the phase error still present, whereas the Fourier type uses a point-source reference.

There are other types of Fourier-transform holograms that use a lens to directly form the Fourier transform. If the distances are appropriately balanced, then the requirement for a point reference source is reduced. In any event, the distinction is in how the quadratic-phase components in the hologram plane are dealt with.

Characteristics of the Reconstructed Image

When one observes a reconstructed image from a hologram, some visual cues are immediately apparent. First, there is noticeable parallax. That is, one object in the image can change in relative position to other objects in the image. Figure 8-7 illustrates this parallax by two different views of the dog and the EE letters. This parallax can be very dramatic in terms of seeing a portion of an object or not seeing it. It is also important in a quantitative sense for measurement. Since relative phase is available, appropriate scales in the image can be used to determine specific properties of the object. This is useful, for example, in double-exposure shock tube holograms where pressure wave measurements are superposed to show the shock wave fronts.

Another dramatic effect noticeable in holograms is the field of view. In a conventional photograph, one focuses a plane onto the photographic film. In a hologram, any plane of the original scene can be brought into focus. The only limitation is the coherence length of the laser used in construction of the hologram. This property is particularly valuable in studying events in media that have three distinct directions, for example, a biological medium. The ability to have ready access to a countable infinity of planes means that the information storage capacity of hologram is very large. In addition, this type of image formation puts a whole new view on microscopy where a limited field on view has been such a problem.

A third effect is often referred to as the keyhole effect. If one peers through a keyhole, the amount of scene observed on the other side depends

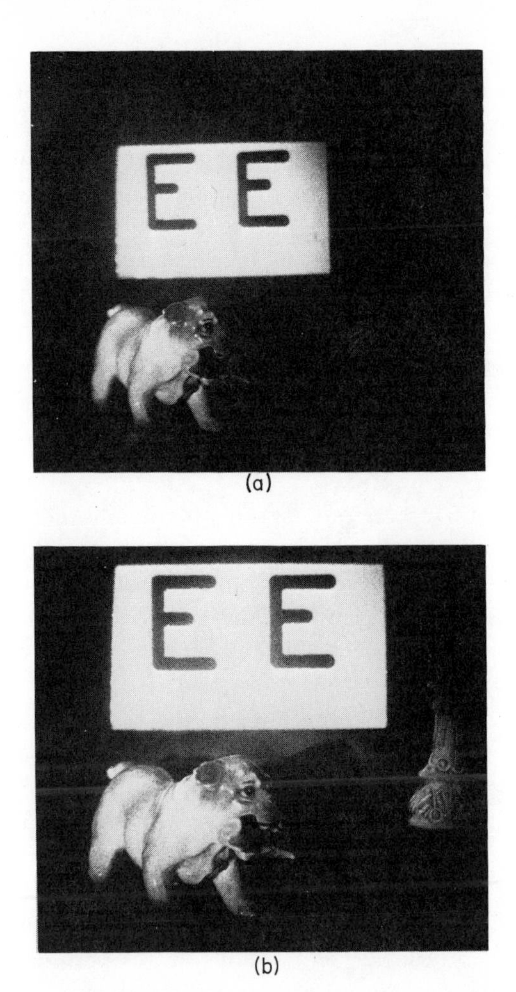

Fig. 8-7. Examples of sideband holograms showing changes in field of view and effect of angle changes.

on how close one is to the hole. In addition, several keyholes in the same door do not provide much additional information, only a change in aspect. A similar effect is obtained by viewing a small portion of a hologram or by viewing the broken pieces of a hologram. Each piece contains information about the whole scene. Only detail and sharpness are improved by larger and larger pieces.

This added feature of redundancy is useful for other applications of holography. It implies a less stringent requirement on the resolution capability of a point on the hologram, since other regions, when added to the reconstruction, will only improve the resolution. There is also the implication of less error-free recording since faults in recording do not destroy detail about a point on the object. Only the sharpness of that detail is reduced.

Other characteristics of the hologram image are demonstrated by reversing and inverting the hologram in the reconstruction setup. These are best understood by using a hologram, since inversions do not necessarily invert the image, only virtual and real images are reversed. Rotations often only shift the images about the reference normal. Combinations of these provide many interesting puzzles for the reader.

Contrast Ratio and Large Dynamic Range

The contrast ratio of an ordinary photograph is determined by the optical density range, the fog level, and the transfer function of the film. Since the fog level is finite and the shoulder of the H–D curve is limited in density to a few optical density units, the contrast ratios are limited. Therefore an equivalent concept of dynamic range is restricted to 40–50 dB at most.

In holography, there is no fog level in the reconstructed image. The fog level of the film affects only the scattering capability of the interferogram. Where no light is scattered, the blackness is very black. The maxima are only limited by the efficiency of the scatterer and the amplitude of the reconstruction beam. Thus the dynamic range is orders of magnitude better, and theoretically the normalized contrast ratio can clearly have values associated with the ideals of zero and unity. This capability of the hologram to have well-defined black or zero regions means that very good optical filtering is possible, and the question simply becomes one of scattering efficiency of the interferogram.

Bandwidth Requirements for Separation

Two cases of hologram image formation have been described. First was the case where all beams were superposed, allowing no clear perception of the reconstructed image, and secondly, where the angle between the reference wave and image wave is sufficient to separate the two conjugate images. In order to make this meaningful, it is necessary to quantify the limits on angle and to show how these limits are related to the spatial information in the recorded image. To do so, Eq. (8-7) can be written as

$$U = (A_1^2 + A_2^2)A_2 \exp(ikr \cos \theta_2) + A_1A_2^2 \exp(ikr \cos \theta_1) + A_1A_2^2 \exp[-ikr(\cos \theta_1 - 2 \cos \theta_2)] \tag{8-9}$$

where

$$\psi_1 = kr \cos \theta_1 \tag{8-10a}$$

and

$$\psi_2 = kr \cos \theta_2 \tag{8-10b}$$

Now since θ_2 is the reference wave angle with respect to the normal and is a known constant, then the concern is whether the variations in θ_1 will be small enough to not overlap through the distortion term bandwidth. The angle θ_1 is just another representation of the spectrum of the input image. One can think of θ_1 as the angular displacement of the spatial-frequency components. If the image content (e.g., a clear transparency) of the input image is low, then the spread on shifted components about θ_1 is very low. If the input image contains considerable fine detail information, then the spread will be large.

A quantitative handle on this can be obtained by taking the Fourier transform of the output and noting that the exponential terms act as shifting terms, displacing the spectrum in the Fourier plane. In Fig. 8-8 a one-dimensional spectrum is shown. Since the distortion terms are products of the amplitude spectrum, they will be convolutions of their own spectrum. Thus the second term on the spectral axis is twice the bandwidth of the amplitude spectrum. The arrow or delta function symbolizes the zero bandwidth of the reference source. The up-shifted and down-shifted terms are directly proportional to input signal spectrum and its corresponding bandwidth.

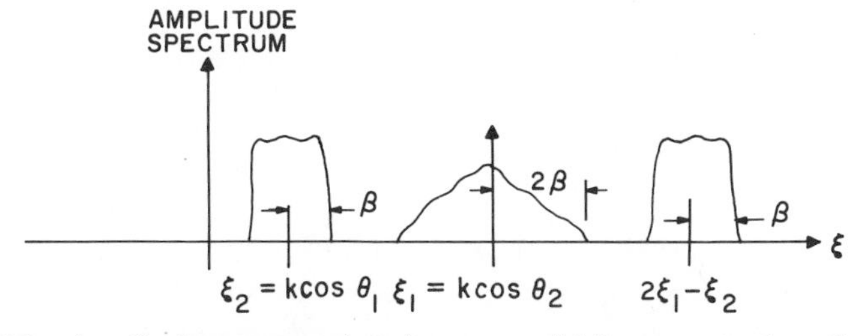

Fig. 8-8. Amplitude spectrum plotted versus spatial frequency to show the sidebands that determine minimum bandwidth constraints.

From this it can be seen that, if the input image has a bandwidth of β, then the minimum angle to keep the images separated is roughly proportional to three times the bandwidth of the input image, or

$$\cos \theta_{2_{\min}} \simeq 3\lambda\beta \tag{8-11}$$

It can also be seen that, if the magnitude of the reference wave is much larger than the amplitude of the scattered object wave, the contribution from the convolution of the amplitude terms can be neglected, and the bandwidth requirement is reduced to

$$\cos \theta_{2_{\min}} \simeq \lambda\beta \tag{8-12}$$

Storage of Multiple Images

Since the information about a particular image is stored in a finite region of the angular spectrum of the hologram output, the carrier can be shifted (angular position of the reference wave) so that several images can be stored on a single hologram. The number of images that can be stored on a single hologram by angular variations is a function of the spatial bandwidth required for each particular image and the bandwidth capability of the interferogram. This latter bandwidth limitation is on the order of 1000 *ll*/mm to 5000 *ll*/mm. The limitation in the spatial bandwidth of the interferogram is primarily related to its ability to efficiently scatter at angles large compared to the hologram normal.

From the previous section on minimum-angle requirements, it is seen that each stored image will require about eight times the bandwidth of the object in order not to have overlap and distortion in the reconstructed waves. The position of the first image is given by

$$\cos \theta_1 = 3\lambda\beta_1 \tag{8-13}$$

where β equals the spatial bandwith of object. Lambda, λ, is the wavelength of the reconstruction wave. The next angular position is given by

$$\cos(\theta_2 + \theta_1) = 4\lambda\beta_1 + 4\lambda\beta_2 \tag{8-14}$$

where β_2 is the spatial bandwidth of the second object wave. This process can be extended to the point that the sum of the angles θ_i is somewhat less than 80° to 85°. At this point the scattering efficiency of the hologram is very poor.

Another method for storing multiple holograms on a single plate is to simply mask off various regions in grid fashion and retrieve the desired image by spatial coordinates of the output illumination. This is much like the cellular storage of a digital computer. The one point that is interesting is that due to the redundancy of information storage in a particular region, there can be overlap at the boundaries of the cells without serious degradation of the image being recovered. This idea will be expanded on in the next section dealing with the use of redundancy in hologram storage.

Reduction in Resolution Requirements through Redundancy

There are several interesting concepts of storage that arise when one considers how a hologram records information about an object over the entire hologram plane. The fact that one can view the object wave in some limited form at almost any sector of the hologram suggests that errors that arise at a particular region will not totally obliterate the object wave or segments thereof. Thus the resolution requirements over a finite cell are not as restricted as in a conventional photograph. Since a composite is formed on total reconstruction, regional imperfections are not as important.

This concept is suggestive of the fact that holographic recording techniques could be used in an analogous manner in digital computer memories. This concept is not new, and efforts are presently underway to apply these concepts. This is particularly attractive since dependence on a particular cell of a discrete memory element is not so important, thus reducing fault tolerance requirements.

Another aspect of the hologram formation might also be applicable to discrete memory techniques. Since the interferogram contains both the potential for viewing the object wave when a reference wave is the reconstructing wave, and also the potential for viewing a reference wave when a replica of the object wave is the reconstructing wave, one could possibly formulate a new technique for digital computer memories following the matched-filter principles. However, the fact that each sector of the hologram contains information about the object wave means that a hologram—matched-filter memory—could be used as a comparative read-only memory where the word size would dictate the resolution and illumination required. Since several object waves could use the same reference wave, considerable overlap could be tolerated still providing adequate readout. Further work in this area remains to be done.

Holographic Interferometry

Because the reconstructed wave from a hologram has the complete amplitude and phase information regarding the object wave, it is possible to consider interference between either the reconstructed wave and the original object wave, the reconstructed wave at $(\lambda_1, \mathbf{r}_1)$ and another reconstructed wave at $(\lambda_2, \mathbf{r}_2)$, or between reconstructed waves that traverse different media to the observer. In any case, the effect is the same. The waves emanating from the same apparent object points have different phase lengths when received at the observer point or plane. This phase difference can be used in a variety of ways to extract information about the object [78, 81–88].

Contour Generation

One of the more interesting uses of this phase difference effect is in the generation of object contours or topographic mapping [74]. These contours can be generated in several different ways. Functionally, however, the intent is the same—to generate a phase difference for each point. These phase differences are observed or recorded by noting that when the phase difference is some multiple of 2π, constructive interference occurs and a fringe is formed. These fringes are related to the actual changes on the object, or the equivalent constant phase points map the contour of the object. To do this pragmatically, a reference plane needs to be defined, and then the changes are referenced to that plane.

Consider for example the setup in Fig. 8-9 where the body surface has either displaced from z_1 to z_2 or the wavelength has shifted from λ_1 to λ_2.

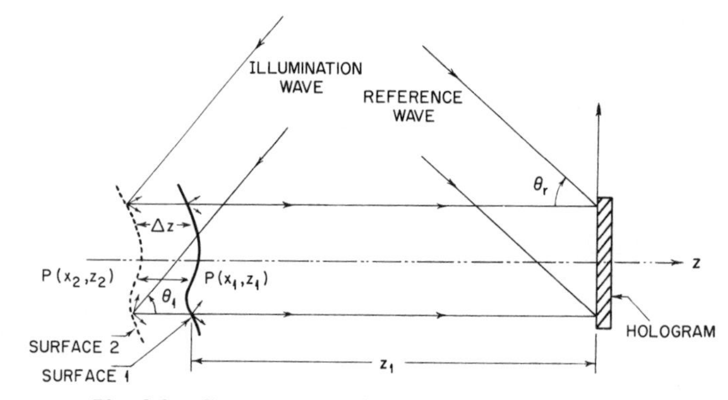

Fig. 8-9. Contour generation—two wavelength method.

In the case that the reconstruction wavelength has shifted, the object upon reconstruction appears to be at the second surface position. That is, the point $P(x_1, z_1)$ has now apparently shifted to point $P(x_2, z_2)$. The amount of shift can be evaluated by equating the electrical lengths from the hologram plane to each point set. The point z_2 can then be solved for as

$$z_2 = z_1(\lambda_1/\lambda_2) \tag{8-15}$$

The lateral shift in the x direction can be calculated noting that the shift in x is due primarily to the difference in the reference wave angle θ_r, reconstruction wave angle θ_c, and the change in z. This can be written as

$$x_2 = x_1 + z_2 \tan\theta_c - z_1 \tan\theta_r \tag{8-16}$$

or using Eq. (8-15)

$$x_2 = x_1 + z_1[(\lambda_1/\lambda_2)\tan\theta_c - \tan\theta_r] \tag{8-17}$$

Now for small angular shifts the tangents can be approximated by their argument, leaving

$$x_2 = x_1 + z_1[\theta_c(\lambda_1/\lambda_2) - \theta_r] \tag{8-18}$$

The lateral shift can be eliminated by adjusting θ_c to $\theta_r(\lambda_2/\lambda_1)$. Equation (8-18) is now in a form that permits evaluation of the change in z.

$$\Delta z = z_2 - z_1 = z_1(\lambda_1/\lambda_2 - 1) = z_1[(\lambda_1 - \lambda_2)/\lambda_2] \tag{8-19}$$

To evaluate the change in height from fringe information, information regarding the change in phase between the surfaces is needed. This is done by subtracting the electrical length $k_1z_1(x)$ from $k_2z_2(x)$ and defining a total phase function $\phi(x)$:

$$\phi(x) = k_2z_2(x) - k_1z_1(x) = 2\pi z_1(x)[(\lambda_1 + \lambda_2)/\lambda_1][(\lambda_1 - \lambda_2)/\lambda_2^2] \tag{8-20}$$

which for small differences in wavelength $(\lambda_1 + \lambda_2)/\lambda_1$ is approximately two, leaving

$$\phi(x) \equiv 4\pi z_1(x)[(\lambda_1 - \lambda_2)/\lambda_2^2] \tag{8-21}$$

This expression can be used to evaluate small equivalent shifts $\delta z_1(x)$ in $z_1(x)$, in terms of incremental changes in phase $\delta\phi(x)$. These increments of phase shift will show up every 2π increment by a fringe formation. Thus,

$$\delta z_1(x) = [\delta\phi(x)/4\pi][\lambda_2^2/(\lambda_1 - \lambda_2)] \tag{8-22}$$

is an equation relating a small height change or contour change on the first surface. The first fringe results when $\delta\phi = 2\pi$, therefore

$$\delta z_1(x)\,|_1 = \tfrac{1}{2}(\lambda_2^2/\Delta\lambda) \tag{8-23}$$

which for two spectral lines in the argon laser system (496.5 nm and 488.0 nm) a $\delta z_1(x)$ of 14 μm results. In some cases, this kind of a detail in contour is too much detail, and thus this contouring method is limited to changes on the order of 1 mm at most.

Contour Generation—Immersion Method

The above description depended on a change in wavelength to achieve the phase change. The same results can be obtained by an equivalent change in wavelength. That is, the dielectric constant can be changed in the medium to effect an equivalent change in λ through changes in k.

In Eq. (8-20) it was implicitly assumed that the free-space wavelength was being considered. Now since η is changing between successive exposures or observations, different paths to a common point on the object need to be considered. The reason that the path changes is that no longer can one employ the same lateral shift control [Eq. (8-18)]. In this case, the angular shifts are controlled by Snell's law. Using this constraint on the angles θ_i, we can evaluate the difference in path length by using Fig. 8-10 and some simple geometry. From Snell's law it is known that

$$\eta_1 \sin\theta_1 = \eta_0 \sin\theta = \eta_2 \sin\theta_2$$

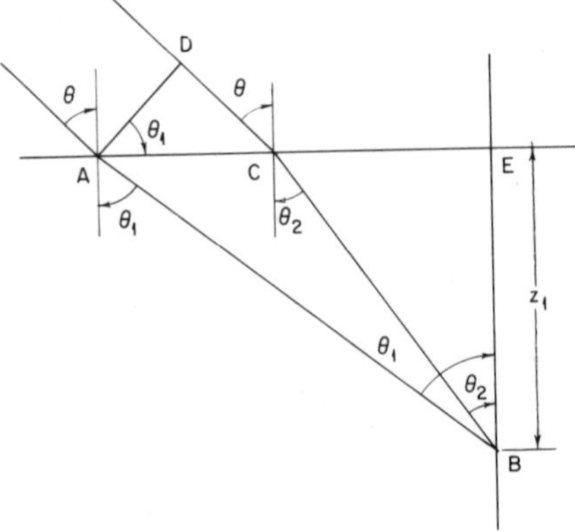

Fig. 8-10. Geometry for immersion method of contour generation.

With this the optical path difference can be written as p, where the plane at AD is the reference point,

$$p = \eta_1(AB + z_1) - \eta_0 DC - \eta_2 CB - \eta_2 z_1 \tag{8-24}$$

Now from trigonometry

$$AE = AB \sin\theta_1 \tag{8-25a}$$

$$CE = CB \sin\theta_2 \tag{8-25b}$$

$$DC = AC \sin\theta_1 \tag{8-25c}$$

$$AC = AE - CE = AB \sin\theta_1 - CB \sin\theta_2 \tag{8-25d}$$

and

$$AB \cos\theta_1 = CB \cos\theta_2 = z_1 \tag{8-25e}$$

Therefore, solving for AB and CB from the last equation, the length of DC can be written as

$$DC = \frac{z_1 \sin\theta_1}{\cos\theta_1} - \frac{z_1 \sin\theta_2}{\cos\theta_2} \tag{8-26}$$

Combining this with Snell's law, p is now written as

$$p = \eta_1 z_1\left(\frac{1 + \cos\theta_1}{\cos\theta_1}\right) - z_1 \frac{\sin^2\theta_1}{\sin\theta_2}\left(\frac{\sin\theta_1}{\cos\theta_1} - \frac{\sin\theta_2}{\cos\theta_2}\right) - \eta_2 z_1\left(\frac{1 + \cos\theta_2}{\cos\theta_2}\right) \tag{8-27}$$

This equation can be rewritten for small angle θ and small perturbations in η as

$$p = \eta_1 z_1(1 + \cos\theta_1) - \eta_2 z_1(1 + \cos\theta_2) \tag{8-28}$$

From this expression the phase difference can be written as

$$\delta\phi(x) = kp = (2\pi/\lambda)z_1(x)[\eta_1(1 + \cos\theta_1) - \eta_2(1 + \cos\theta_2)] \tag{8-29}$$

which for an incremental phase shift of 2π, the first fringe forms, leaving

$$\Delta z_1(x) = \frac{\lambda}{\eta_1(1 + \cos\theta_1) - \eta_2(1 + \cos\theta_2)} \tag{8-30}$$

This expression is quite important since it suggests that to resolve Δz on the order of 15 to 100 μm with the two wavelength method, changes in η (i.e., $\Delta\eta = \eta_1 - \eta_2$) on the order of 2×10^{-2} to 3×10^{-3} have to be measured. This order of change in η is the same order of magnitude seen in gases when pressures are changed just a few atmospheres. It would seem that the two methods could be used together to measure pressure changes in a closed chamber, particularly in the case of liquids.

Differential Holograms—Strain Measurement

If Eq. (8-20) is examined from the point of view of actually deforming the body, such as under stress, then the fringe information can be used to measure the strain. This technique can be used to accurately measure strain at a single wavelength [81, 83].

In another sense it can all be used to measure the deviations of machined parts from a master layout. In the case of precisely machined parts, this would take the place of the "machinist flat" which uses Newton rings for

Fig. 8-11. Doubly exposed hologram showing strain effects.

surface deformation measurements. This technique, however, suffers from the fact that unless the viewing is done normal to the body, the fringes do not form on the surface. Therefore, a method of translating the deformation data to the surface of the body is necessary. For nonnormal viewing, the displacement data are of little value. Figure 8-11 shows such a body under stress. This is a hologram of a deflected cantilever beam. The fringes are, however, quite visible, and the boundary condition of zero displacement is evident at the fixed end.

This method has the advantage that for bodies with a complex anisotropic Young's modulus, the components of the modulus tensor can in principle be measured. This is particularly valuable in biological structures such as bones where present plastic modeling techniques are inadequate.

Differential Holograms—Vibrational Analysis

The technique of displacement measurement described in the previous section has been successfully applied to vibrational mode measurements [84]. Basically one makes a hologram of a device, such as a transducer, when it is not vibrating, and then compares this nonvibrating image with the vibrating image. If the vibrating image has stationary nodes in the pattern, the fringe pattern will clearly show these points. In addition, average values of displacement will appear through the fringe pattern and clearly show the modal behavior of the vibrational modes. This technique is providing an experimental method for testing instrument and transducer designs that have previously evaded actual design testing. In the case of musical instruments, these patterns were a first in visualizing actual mode vibrations.

Volume Effects—Bragg Angle

One of the physical parameters of a hologram that has been glossed over is the thickness of the emulsion. It has been implicitly assumed that the emulsion is thin and that multiple layers of interference have not been considered. This becomes particularly important when it is noted that emulsion thicknesses of the photographic plates used in holography vary from 5 to 15 μm. Thus if visible radiation ($\lambda \sim 0.6$ μm) is used, then several wavelengths are traversed in an emulsion layer.

Since any scattering object wave can be decomposed into a Fourier spectrum of plane waves, the analysis will be restricted to the interference of two plane waves. If it is assumed that two interfering plane waves intersect the photographic plate at equal angles, $\theta/2$, with the normal, then the geometry shown in Fig. 8-12 of the interferogram and the corresponding planes of deposited silver is an adequate model. The dotted lines correspond to planes of zero phase or planes that are 2π radians apart. Therefore, these planes are separated a distance λ. Wherever these dotted planes intersect, a vertical plane of constructive interference is obtained.

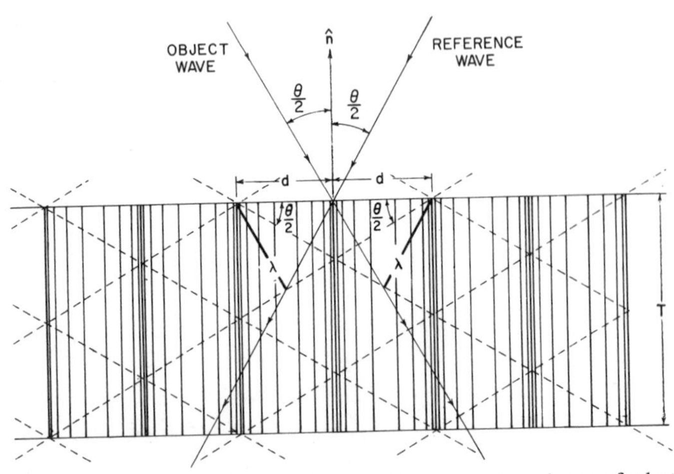

Fig. 8-12. Geometry for volume hologram effects showing planes of photographic density in an emulsion of thickness T.

Thus under development, the vertical planes of constructive interference will have heavy deposits of silver. The separation of the vertical planes or spacing of the interferometer can be calculated from simple geometry, using

$$2d \sin(\theta/2) = \lambda \tag{8-31}$$

The problem associated with these vertical layers arises when one attempts to get efficient reconstruction. In Fig. 8-13 a simplified picture of the layer with the interferometric planes as solid lines is shown with reconstruction wave at an angle ϕ to the vertical plane. If these vertical planes are assumed to be partially reflecting mirrors, then a model for finding the most efficient reconstruction is obtained. It is quite easy to see that maximum brightness

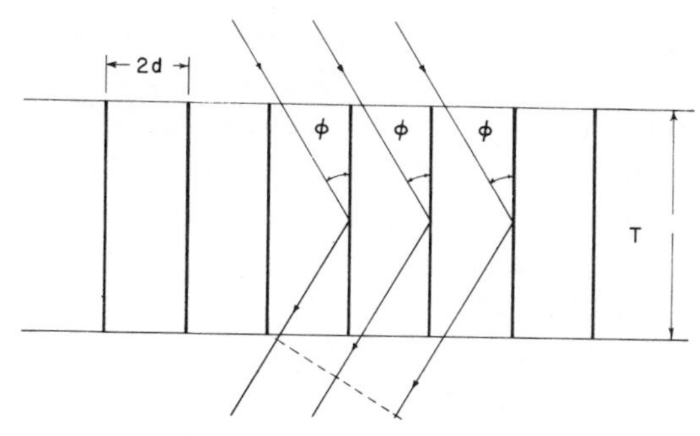

Fig. 8-13. Reconstruction geometry in volume holograms showing planes of deposited silver.

will occur when each of the waves emanating from the partially reflecting mirrors is in phase. This is the so-called Bragg condition. In this geometry it can be written as

$$2d \sin \phi = \pm\lambda \tag{8-32}$$

since each successive wave front is shifted 2π radians in phase. Thus the reconstructing wave has to satisfy the condition that

$$\phi = \sin^{-1}(\lambda/2d) = \pm\theta/2 \tag{8-33}$$

or some π multiple of the angles, thus

$$\phi = \begin{cases} \pm\theta/2 \\ \pm(\pi - \theta/2) \end{cases} \tag{8-34}$$

Now the two possible solutions of this equation correspond physically to the virtual and real images of the reconstructed object wave. This Bragg effect limits the viewing angle and illumination characteristics of volume holograms and imposes some undesired constraints on the illumination.

Since Bragg angle effects are wavelength dependent, it is clear that given angles of illumination will give frequency selective effects. Thus properly configured holograms can serve as white light holograms since they will self-select the proper wavelengths for reconstruction. Thus some researchers have been able to make white-light or multicolor imaging holograms [90] using this effect.

The one remaining question is therefore when one can consider a hologram as a thin hologram or a thick hologram. That is, when do surface effects dominate over the volume effects described above? The answer lies in a comparison of the thickness T and the interferometer spacing $2d$. If it can be shown that the angles and thickness are such that T is much less than $2d$, then the hologram should be considered thin or surface effect dominated. If T is much greater than $2d$, then it is a thick hologram, and volume effects should be considered dominant. In the intermediate range, other considerations will be important. Typical of these is the number of wavelengths involved, the total spread of angles involved in describing the object wave, and the need for maintaining high spatial-frequency information.

Use of a Hologram as a Complex Filter Element

Up to this point, holograms have been described in terms of viewing a three-dimensional image and discussing the inherent properties of such an image. There is, however, an alternate point of view, and that is to utilize the scattered reconstruction wave as an element in or input to a coherent optical processor. As has been noted many times throughout this chapter, the uniqueness of the holographic image or reconstructed wave is that it contains both the amplitude and the phase corresponding to the original object. This is quite an advantage since generally the phase information is very difficult to obtain, particularly if registry with the amplitude information is important. The most obvious example is the equivalent situation in the matched-filter problem. Here the signal is mixed with its complex conjugate (the other twin image) to form the reference wave output. This reference wave output is generally a point since a spherical wave was used in construction.

Holographic Lens

Another example is the formation of the holographic lens [86]. Here the object wave is the wave from a point source, and upon reconstruction a focused wave is formed in a plane. This technique, shown in Fig. 8-14, can be used to form a lens that can later be modulated by other input formats.

To analytically describe the process, the field in the hologram plane is represented by two terms $P(x, y)$ and $\alpha(x, y)$, where

$$P(x, y) = P \exp[-i(k/2f)(x^2 + y^2)] \tag{8-35}$$

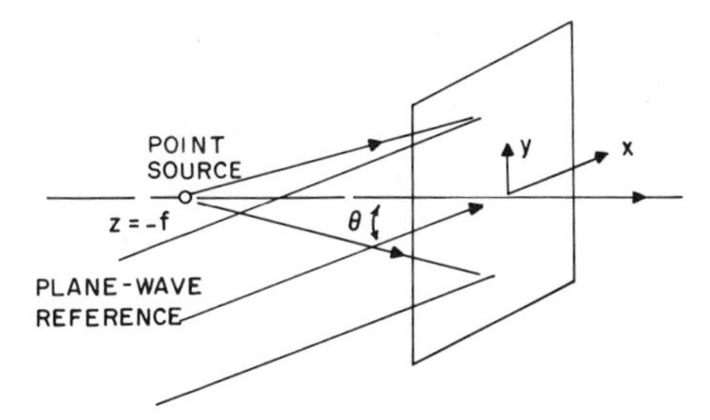

Fig. 8-14. Setup for making a holographic lens using a point source and a plane wave reference beam.

and

$$\alpha(x, y) = A \exp[-iky \sin \theta] \tag{8-36}$$

correspond to the point source and the reference plane wave, respectively. If the knee of the H–D curve is used, the amplitude transmission function of the hologram can be expressed [87] as

$$t(x, y) = \sum_{n=-\infty}^{+\infty} t_n(x, y) = \sum_{n=-\infty}^{\infty} (-i)^n J_n(2cPAT) \exp[i(kn/2f)(x^2 + y^2)] \times \exp[inky \sin \theta] \tag{8-37}$$

where c and T are constants associated with the photographic process. Thus if an appropriate plane wave illuminates the hologram, a series of output waves will each occur at different angles in θ. If the $t_{-1}(\ )$ term is specifically examined when a conjugate plane wave (symmetric to the hologram normal) is used for reconstruction, the following is obtained for the output:

$$0(x, y) = \mathscr{A}^*(x, y) t_{-1}(x, y) \tag{8-38}$$

where

$$\mathscr{A}^*(x, y) = A \exp[iky \sin \theta] \tag{8-39}$$

This reduces to

$$0(x, y) = \mathscr{A}^* t_{-1} = iAJ_{-1}(2cPAT) \exp[i(k/2f)(x^2 + y^2)] \tag{8-40}$$

which is just a focusing wave having an apparent focal point at $z = -f$. Notice that the $\mathscr{A}(x, y)$ used in the reconstruction process could be another input signal format; thus $t_{-1}(x, y)$ is just a conventional lens function.

Figures 8-15 and 8-16 show a case where a grating function was used for $\mathscr{A}(x, y)$. The output is then the classical diffraction pattern of a grating. It has been shown [86] that the quality of such lenses can be very good as indicated by classical Ronchi lens tests.

This technique can be further extended to use two identical hologram lenses placed in conjugate orientation as shown in Fig. 8-17. This setup is used to minimize the effects of the aberrations in the lens due to the off-axis formation. This technique also has the advantage of bringing the optical axes throughout the system into parallel alignment.

Fig. 8-15. Fourier transform of a horizontal grating using a holographically generated lens. (a) Input grating. (b) Output transform.

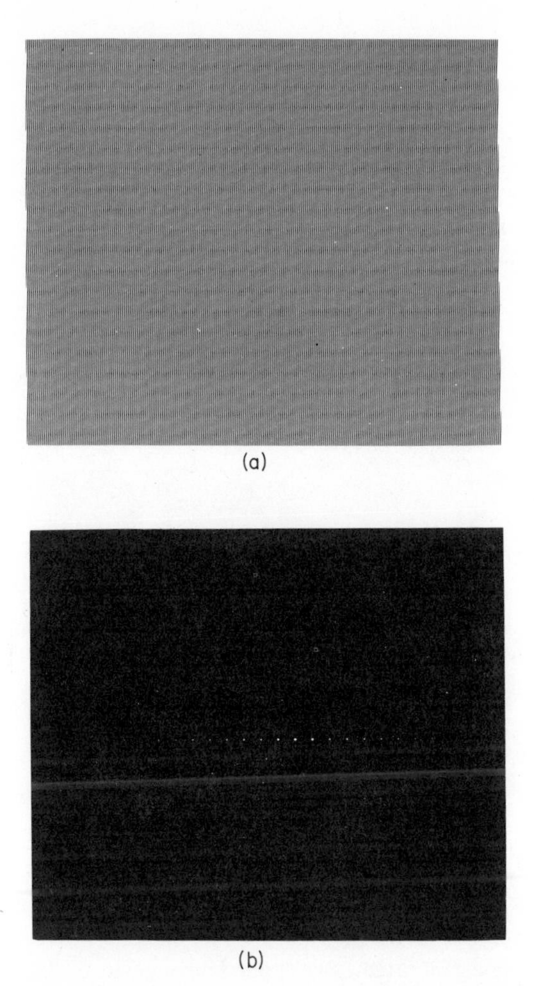

Fig. 8-16. Fourier transform of vertical grating using a holographically generated lens. (a) Input grating. (b) Output transform.

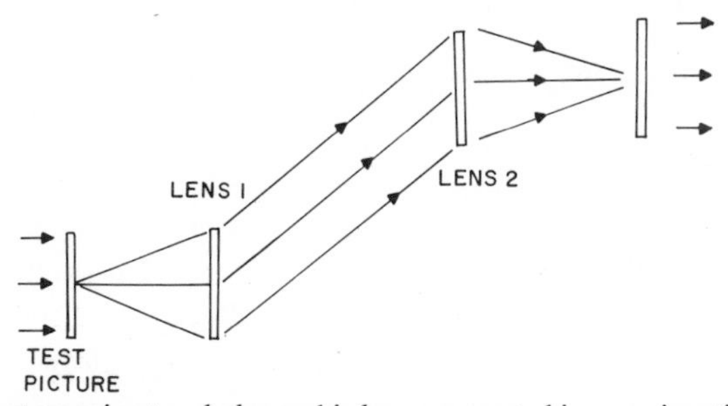

Fig. 8-17. A system using two holographic lenses mounted in a conjugating configuration to correct for inherent lens aberration and coma.

Aberration Correction

These same techniques can be used [87] to correct the aberrations and other phase distortions of a given lens. In this case, the lens to be corrected is used as the input object in construction of the hologram. Upon reconstruction the conjugate image from the hologram of the object lens is mixed with the lens output to cancel the phase errors or aberrations. In Fig. 8-18 details of one setup are shown.

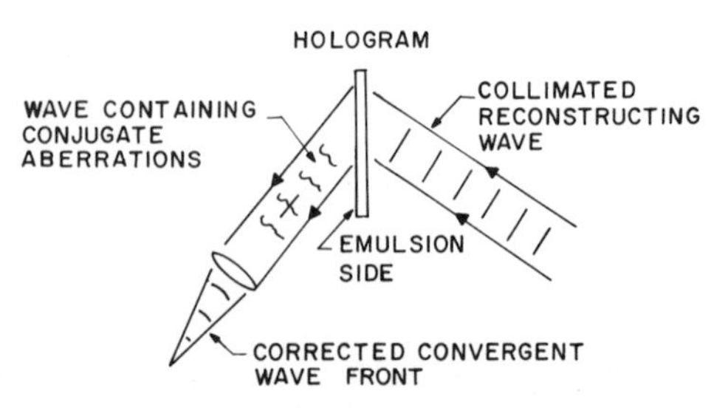

Fig. 8-18. A system setup for correcting the inherent aberration of a lens through the use of a hologram of the lens.

Use of a Hologram as a Generalized Processor Element

So far specific examples of how a hologram can create a desired complex wave have been described. There is nothing in the formation to restrict the generalization of this concept. Using the holographic techniques described above, the filters necessary to do almost any linear mathematical operation can be generated. This is particularly true of the class of operations described by a Fredholm integral equation. That is, if $\mathscr{L}\{f(x, y)\}$ is some mathematical operation representable by

$$\mathscr{L}\{f(x, y)\} = (\,) \iint f(x', y')K(x, y; x', y')\, dx'\, dy' \tag{8-41}$$

then in principle the necessary kernel $K(x, y; x', y')$ can be generated through the use of hologram filters. This brief statement is not to suggest that this would be trivial, but rather to suggest that conceptually a means of generating the required filters has been described. In Chapter V several examples of where this capability would be very desirable were shown. It would appear that research in this direction will yield further examples of processors developed with these holographic techniques [88].

Problems

1. What additional information is provided in a hologram that is not provided in a conventional photograph? How is this done?

2. What is the essential difference between the Vander Lugt matched filter and a hologram?

3. How does the dynamic range of a holographic image compare to a photographic image?

4. What is the minimum angle required for making a hologram when the upper bandlimit of image detail is 2000 *ll*/mm?

5. If the gamma, γ, of a film were fixed at $\gamma = 3$, what constraints would this impose on making holograms with this film?

6. What is the smallest length increment that can be detected using contour generation techniques and $\lambda = 488.0$ nm and 514.5 nm?

Chapter IX

PARTIAL COHERENCE

Introduction

So far the development of optical concepts has focused on fields and intensities that are derived from monochromatic sources with deterministic spatial phase variation. It has been assumed that measurements about the field amplitudes could be done by measuring the intensity. It was further asserted that the optical phase information could be obtained by interferometric means.

In general, however, precise amplitude and phase parameters are difficult to obtain directly. This information is generally obtained from fringe formations. Thus, it is natural to wonder how much the fringe formations and intensity patterns are affected by variations in frequency and phase. Early work on the effects of source size and fluctuations in source parameters was done by Verdet (1869) [91], Michelson and others [92–99]. Discussions on the evolution of partial coherence theoretical concepts are given in Born and Wolf [11], Beran and Parrent [59], and Mandel and Wolf [100]. The basic theory seems to be reasonably well developed; however, its application to and impact on optical system design are not so well defined.

Designers and researchers in electromagnetic and optical systems have, except in the sixties and seventies, restricted analytical design methods to consideration of two basic types of sources [100]: those that have well-defined monochromaticity and phase variations (coherent), or to sources of broad spectral width and random spatial phase variation (incoherent). The study of partial coherence is then an attempt to understand and design systems when neither of these extreme conditions apply.

Fringes and Monochromaticity

Since the observables in optical systems are second-order measures like intensity, the first inquiries were directed at studying the changes in interference effects as source parameters and simple imaging systems were varied. The simplest system to examine is the two-slit Young's interferometer (see Chapter VII) when nonideal sources are used to excite the system. The formalism necessary for examining this system was provided in the work of Chapter II on two-aperture radiating systems.

The far field from such a two-aperture system is basically represented by two multiplicative spatial terms. One is the aperture factor, which is determined by the size of the apertures. The second is the array factor, which is determined primarily by the aperture separation. Neglecting the linear phase term and other constants, we can write the output field as

$$u(x', y') = [\ ]\left|\cos\left(\frac{kx'd}{l}\right)\right|\left[\frac{\sin(kx'a/l)}{(kx'a/l)}\ \frac{\sin(ky'b/l)}{(ky'b/l)}\right] \tag{9-1}$$

where d characterizes the aperture separation, a and b are the aperture size parameters, and l is the path length. If ka and kb are very small compared to π, the first-order off-axis variations in the output field will be determined by the array factor, $\cos(kx'd/l)$. This is just the case in the two-slit Young's experiment. Similar results are obtained with two small holes, except that the aperture symmetries are circular. For the moment only the one-dimensional two-slit problem will be examined.

If everything is ideal, the first point of destructive interference occurs at the first zero of the array factor, i.e.,

$$kx_d'd/l = \pm\pi/2 \tag{9-2a}$$

or

$$kx_d' = \pm(l/d)(\pi/2) \tag{9-2b}$$

For most experimental problems the ratio l/d is a number on the order of ten to one hundred. In terms of the size of k, it is small. The key question for nonmonocromatic sources of small-frequency deviation is what happens to this null position as frequency variations are considered. To explore this, rewrite x_d' in terms $\omega \pm \delta\omega$ and the velocity of light† v in the medium

$$x_d' = \frac{(l\pi/2d)v}{\omega(1 \pm \delta\omega/\omega)} \tag{9-3}$$

† The velocity of light in a general medium characterized by the permittivity ε and permeability μ is $1/(\varepsilon\mu)^{1/2}$, which can be written in terms of the free-space velocity c as c/η, where η is the refractive index $(\varepsilon/\varepsilon_0)^{1/2}$, and μ equals μ_0.

which, for $\delta\omega \ll \omega$, becomes

$$x_d' = (l\pi/2d)(v/\omega)(1 \mp \delta\omega/\omega) \tag{9-4}$$

If ω represents the mean frequency in the medium and λ the mean wavelength, then

$$x_d' = (l\lambda/4d)(1 \mp \delta\omega/\omega) = x_{do}' \mp \delta x_d' \tag{9-5}$$

where

$$\delta x_d' = (l\lambda/4d)(\delta\omega/\omega) \tag{9-6}$$

The shifts in null are on the order of $\delta\omega/\omega$, modified by the ratio of $(l\lambda/4d)$. The net effect of this ambiguity is that where once a clear null was evident, now only a relative null will appear. Pictorially this can be represented by the pictures shown in Fig. 9-1.

Fringes and Phase Perturbations

A similar analysis would show that if the relative phase in two apertures were to fluctuate, Eq. (9-2a) can be rewritten as

$$\frac{kx_d'd}{l} + \delta\theta = \frac{\pi}{2} \tag{9-7}$$

where

$$x_d' = \left(\frac{l\lambda}{4d}\right)\left(1 \mp \frac{\delta\theta}{\pi/2}\right) = x_{do}' \mp \delta x_d' \tag{9-8}$$

and

$$\delta x_d' = \left(\frac{l\lambda}{4d}\right)\left(\frac{\delta\theta}{\pi/2}\right) \tag{9-9}$$

This result is, however, like Eq. (9-6) in that the null position shifts are on the order of the phase fluctuations normalized to $\pi/2$. Thus Fig. 9-1 is also a description of what happens in the case of phase perturbation across the wave front.

It is obvious that Fig. 9-1(a) describes an ideal situation (coherent) and 9-1(d) describes the worst situation (incoherent). Figure 9-1(b) and 9-1(c) are something intermediate. The quantitative description of these less specific cases will be developed in later sections.

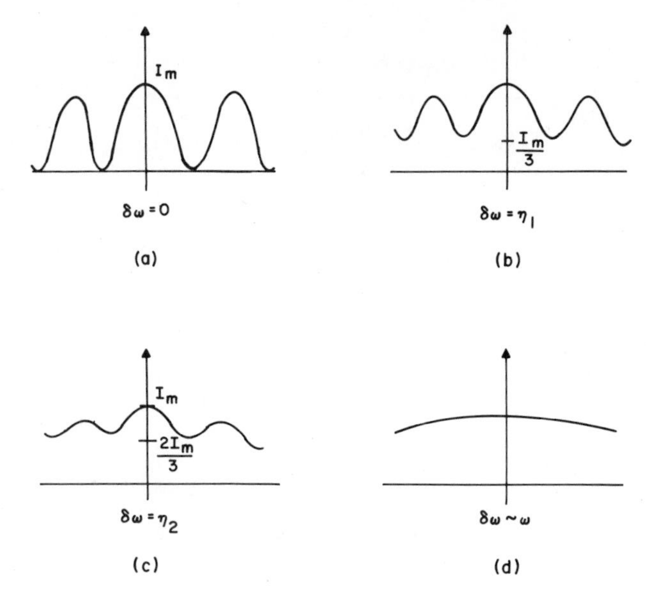

Fig. 9-1. Fringe changes and associated visibility for changes in monochromaticity or phase.

Visibility

It is desired at this point to find some way of measuring or quantifying the differences between these various cases. Study of Fig. 9-1 leads one to realize that it has to do with the visibility of the fringes. That is, the differences between the adjacent maxima and minima of the intensity. Traditionally, the quantity used to quantify the differences in fringe formation is the visibility function V, defined as

$$V = \left| \frac{I_{\max} - I_{\min}}{I_{\max} + I_{\min}} \right| \tag{9-10}$$

This function is a good measure for the various cases given in Fig. 9-1. Using Eq. (9-10), we show in Table 9-1 the calculations for each case in Fig. 9-1. From these cases it is clear that the coherent case is associated with V equal to unity, and the incoherent case is V equal to zero. Thus V by definition is restricted to values between zero and one. The states in between signify some state of partial coherence. In a later section, the fringe visibility will be associated with the concept of the degree of coherence γ. The fringe visibility model is an effective way of tying the general concept of coherence with effects that are measurable.

Table 9-1

Calculation and Designation of Partially Coherent Fringes Shown in Fig. 9-1

9-1(a)	$V = \dfrac{I_m - 0}{I_m + 0} = 1$	coherent
9-1(b)	$V = \dfrac{I_m - I_m/3}{I_m + I_m/3} = 0.5$	partially coherent
9-1(c)	$V = \dfrac{I_m - 2I_m/3}{I_m + 2I_m/3} = 0.2$	partially coherent
9-1(d)	$V = \dfrac{I_m - I_m}{I_m + I_m} = 0$	incoherent
	$V = \left\lvert \dfrac{I_{\max} - I_{\min}}{I_{\max} + I_{\min}} \right\rvert$	

Mutual Coherence Function

A firmer foundation can be established by defining a basic function that interrelates the fields from each aperture at every point in the space. The relation between the field functions must somehow measure their coherence properties. The quantity that has been traditionally defined is the mutual coherence function $\Gamma(\ )$ given by

$$\Gamma_{12}(\tau) = \Gamma(\mathbf{x}_1, \mathbf{x}_2, \tau) = \langle V_1(\mathbf{x}_1, t + \tau) V_2^*(\mathbf{x}_2, t) \rangle \tag{9-11}$$

The brackets, $\langle\ \rangle$, mean a long-time average, i.e.,

$$\langle u(t) \rangle = \lim_{T \to \infty} \frac{1}{2T} \int_{-T}^{T} u(t)\, dt \tag{9-12}$$

The quantities, $V(\mathbf{x}_1, t_1)$ and $V(\mathbf{x}_2, t_2)$ are the complex field disturbances (analytic signals [59]) at the points $\mathbf{x}_1$ and $\mathbf{x}_2$ and at times t_1 and t_2. In this case, a property of stationarity in time is being assumed, i.e.,

$$t_1 - t_2 = \tau = \Delta l / c \tag{9-13}$$

where Δl is the path length difference between the two observation points and the source point. The property of stationarity has to do primarily with the fact that it does not matter in an absolute time sense when the time samples were obtained, rather what is the time difference between samples. It is important to remember that any sample of the field disturbance is of finite length and the concern is how long the sample is and when it was obtained in a relative sense. Note that when $\mathbf{x}_1 = \mathbf{x}_2$, $\Gamma_{12}(\tau)$ reduces to the self-intensity $I(\mathbf{x}_1)$, which is something directly observable or measurable.

In the case that $\mathbf{x}_1 \neq \mathbf{x}_2$, the question of what is observable becomes a problem. Because direct measurement of field variables is not obtainable, one is forced to use inferred measures at other observation points. In a strict sense, an interferometer is being formed to gain insight into an observable that cannot be obtained directly. The classical Young's two-slit interferometer, shown in Fig. 9-2, is such an interferometer. From the measurement of the intensity at P and other points along a plane parallel to the plane of the slits, the mutual coherence of the disturbances at $\mathbf{x}_1$ and $\mathbf{x}_2$ can be inferred. To develop the details of this measurement, some specific parameters have to be defined and measured.

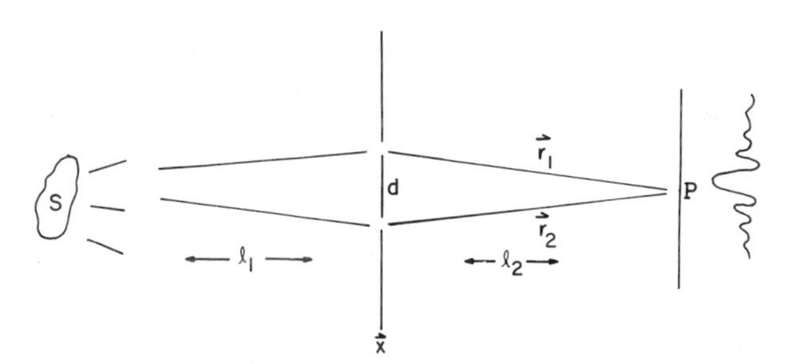

Fig. 9-2. Young's interferometer for measurement of the mutual coherence function of the source S.

Complex Degree of Coherence

How these measurements of P are made and used can best be described by defining the complex degree of coherence

$$\gamma_{12}(\tau) = \frac{\Gamma_{12}(\tau)}{[\Gamma_{11}(0)\Gamma_{22}(0)]^{1/2}} \tag{9-14}$$

which is just a normalized form of the mutual coherence function. The normalizing parameters $\Gamma_{11}(0)$ and $\Gamma_{22}(0)$ are obtained by first blocking slit number two and measuring the self-intensity due to slit number one, $I_1(\mathrm{P})$ or $\Gamma_{11}(0)$, and then blocking slit number one and measuring $I_2(\mathrm{P})$ or $\Gamma_{22}(0)$. These self-intensities are straightforward measurements and serve primarily to bound the magnitude of the complex degree of coherence by zero and one, the same as the fringe visibility function.

Next, the measurements made at P need to be related to the disturbances at $\mathbf{x}_1$ and $\mathbf{x}_2$. In Chapter II the field at P due to slits (or apertures) one and two was related through the Rayleigh–Sommerfeld integral equation and can be written as

$$V_p(t) = \left(\frac{-j}{\lambda z}\right) \iint_{s_1} V(x, y, t) e^{-jkR_1}\, dx\, dy + \left(\frac{-j}{\lambda z}\right) \iint_{s_2} V(x, y, t) e^{jkR_2}\, dx\, dy \tag{9-15}$$

This representation can be simplified by using a linear operator notation, i.e.,

$$V_p(t) = L_1 V_1 + L_2 V_2 \tag{9-16}$$

where

$$L_i = \left(-\frac{j}{\lambda}\right) \iint_{s_i} (\,)\, \frac{e^{jkR_i}}{R_i}\, dx\, dy \tag{9-17}$$

This notational scheme allows a simpler manipulation of the operations called for in Eq. (9-11). Substitution of Eq. (9-16) into Eq. (9-11) and expanding yield

$$\langle V_p(t_1) V_p{}^*(t_2) \rangle = \langle (L_1 V_1 + L_2 V_2)(L_1 V_1 + L_2 V_2)^* \rangle \tag{9-18a}$$

$$= \langle L_1 L_1{}^* V_1(t_1) V_1{}^*(t_1) \rangle + \langle L_2 L_2{}^* V_2(t_2) V_2{}^*(t_2) \rangle + \langle L_1 L_2{}^* V_1(t_1) V_2{}^*(t_2) \rangle + \langle L_1{}^* L_2 V_1{}^*(t_1) V_2(t_2) \rangle \tag{9-18b}$$

There are inherent assumptions in the explicit form of the mutual coherence function given in Eq. (9-18). One is that the radiation is quasi-monochromatic. By this, it is meant that

$$\Delta f \ll f_0 \tag{9-19}$$

Further, it is assumed that the path difference Δl is very much smaller than the wavelength fluctuations $c/\Delta f$.

Equation (9-18b) uses t_1 and t_2 to denote the times at point p corresponding to the time at the aperture points $\mathbf{x}_1$ and $\mathbf{x}_2$. These times can be explicitly written in terms of the path lengths as

$$t_1 = t - r_1/c \tag{9-20a}$$

$$t_2 = t - r_2/c \tag{9-20b}$$

Substitution of Eqs. (9-20) into Eq. (9-18b) yields

$$\begin{aligned} I(\mathrm{P}) = &\left\langle L_1 L_1{}^* V_1\left(t - \frac{r_1}{c}\right) V_1{}^*\left(t - \frac{r_1}{c}\right)\right\rangle \\ &+ \left\langle L_2 L_2{}^* V_2\left(t - \frac{r_2}{c}\right) V_2{}^*\left(t - \frac{r_2}{c}\right)\right\rangle \\ &+ \left\langle L_1 L_2{}^* V_1\left(t - \frac{r_1}{c}\right) V_2{}^*\left(t - \frac{r_2}{c}\right)\right\rangle \\ &+ \left\langle L_1 L_2{}^* V_1\left(t - \frac{r_1}{c}\right) V_2{}^*\left(t - \frac{r_2}{c}\right)\right\rangle^* \end{aligned} \tag{9-21}$$

which can be simplified to

$$I(\mathrm{P}) = I_1(\mathrm{P}) + I_2(\mathrm{P}) + 2\,\mathrm{Re}\{L_1 L_2{}^* \Gamma_{12}(\tau)\} \tag{9-22}$$

These terms have specific physical meaning and can be associated with the measurements at P described earlier for total intensity and the self-intensities. The last term in Eq. (9-22) is the real part of the propagated mutual coherence function. The proof that the mutual coherence function propagates is given in a later section.

The problem now reduces to examining the real part of $L_1 L_2{}^* \Gamma_{12}(\tau)$. Since this is a complex quantity, it can be represented in a polar form as having magnitude and complex phase as follows:

$$L_1 L_2{}^* \Gamma_{12}(\tau) = L_1 L_2{}^* \mid \Gamma_{12}(0) \mid e^{j\theta_{12}} \tag{9-23}$$

where θ_{12} has incorporated in it the phase variation due to the time delay τ and the possibility that the apertures may differ in relative phase by the amount β_{12}. This can be expressed as

$$\theta_{12} = \omega(r_1/c - r_2/c) + \beta_{12} \tag{9-24}$$

or using the average wavelength λ,

$$\theta_{12} = (2\pi/\lambda)(r_1 - r_2) + \beta_{12} \tag{9-25}$$

One other point is that the magnitude of the mutual coherence function is referenced to the point of symmetry between the apertures, which corresponds to zero time delay. This is purposely done to allow the introduction of the complex degree of coherence function defined in Eq. (9-14).

Using the notion that Eq. (9-23) represents measurements made in plane P about the coherence function in plane **x**, then substitution of Eq. (9-14) in Eq. (9-23) yields

$$L_1 L_2^* \Gamma_{12}(\tau) = [I_1(\mathrm{P}) I_2(\mathrm{P})]^{1/2} \mid \gamma_{12} \mid e^{i\theta_{12}} \tag{9-26}$$

which reduces Eq. (9-22) to

$$I(\mathrm{P}) = I_1 + I_2 + 2(I_1 I_2)^{1/2} \operatorname{Re}\{\mid \gamma_{12} \mid e^{i\theta_{12}}\} \tag{9-27}$$

If the Euler relation $e^{ix} = \cos x + i \sin x$ is used, Eq. (9-27) becomes

$$I(\mathrm{P}) = I_1 + I_2 + 2(I_1 I_2)^{1/2} \mid \gamma_{12} \mid \cos \theta_{12} \tag{9-28}$$

Having the result in this form is convenient when comparisons to two well-known results are desired.

First, the coherent case where $\mid \gamma_{12} \mid$ is unity gives the same result as the coherent analysis of the Young's interferometer derived in Chapter VII,

$$I_{\mathrm{P}} = I_1 + I_2 + 2(I_1 I_2)^{1/2} \cos \theta_{12} \tag{9-29}$$

Second, the inherent case represented by $\mid \gamma_{12} \mid$ equal to zero corresponds to the simple result of intensity superposition,

$$I_{\mathrm{P}} = I_1 + I_2 \tag{9-30}$$

It is this result that has been used classically to represent optical systems as linear in intensity. This linear intensity theory was used by many early investigators to do the first frequency analysis of optical systems [24].

Measurement of the Degree of Coherence

Equation (9-28) can be used to form a basis for measurement of $\mid \gamma_{12} \mid$. If system parameters can be adjusted such that as there are points in the output plane where θ_{12} goes through values of $2n\pi$ ($n = 0, 1, 2, \ldots$), then a maximum value in intensity, I_p, can be obtained

$$I_{p\,\max} = I_1 + I_2 + 2(I_1 I_2)^{1/2} \mid \gamma_{12} \mid \tag{9-31}$$

If parameters were shifted so that θ_{12} was $(2n + 1)\pi$ $(n = 0, 1, 2, \ldots)$, then a minimum value of intensity would be obtained

$$I_{p\,\min} = I_1 + I_2 - 2(I_1 I_2)^{1/2} \mid \gamma_{12} \mid \tag{9-32}$$

These two extremes do indeed occur in the fringe pattern. Thus, if the apparatus is adjusted to give equal self-intensities, then the magnitude of γ_{12} can be solved for, yielding

$$\mid \gamma_{12} \mid = \frac{I_{p\,\max} - I_{p\,\min}}{I_{p\,\max} + I_{p\,\min}} \tag{9-33}$$

This result is very similar to the fringe visibility function, Eq. (9-10), with the values of $\mid \gamma_{12} \mid$ restricted to

$$0 \leq \mid \gamma_{12} \mid \leq 1 \tag{9-34}$$

A direct coupling has now been made between fringe information and its boundedness, and the measurement of incoherent and coherent states with the complex degree of coherence function. Thus a rather simple measure of the coherence concept has been obtained.

Separation of Spatial and Temporal Effects

The basis for separating the spatial and temporal effects of coherence follows from the definition of $\gamma_{12}(\tau)$ as it is decomposed into a magnitude $\mid \gamma_{12}(0) \mid$ and phase term $\theta_{12}(\tau)$. It becomes apparent when examining Eq. (9-14) and Eq. (9-26) that

$$\gamma_{12}(\tau) = \mid \gamma_{12}(0) \mid e^{i\theta_{12}(\tau)} \tag{9-35}$$

which is consistent with the definition of terms given in Eqs. (9-23) and (9-24). The modulus of Eq. (9-35) represents the spatial effects that arise from relative phase differences between the apertures. The argument of Eq. (9-35) is a measure of the temporal effects associated with frequency and phase fluctuations in the measured fringe pattern. Thus the choice of using the point of symmetry in the P plane for defining $\mid \gamma_{12}(0) \mid$ has important physical meaning.

This assertion can be explored further by investigating the effect of varying the slit separation d. If the slit spacing is varied over a range of d, fringe patterns in the P plane will change in a manner similar to that shown

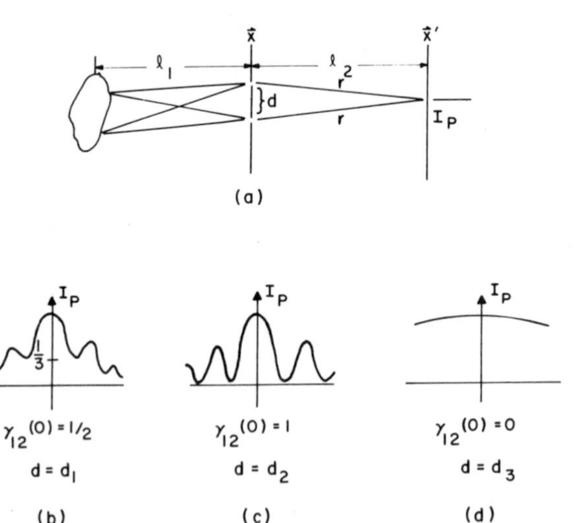

Fig. 9-3. System for measuring degree of coherence changes when sampling aperture separation is varied. (a) Basic system. (b) Intensity profile for $d = d_1$ and $\gamma_{12}(0) = \frac{1}{2}$. (c) Intensity profile for $d = d_2$ and $\gamma_{12}(0) = 1$. (d) Intensity profile for $d = d_3$ and $\gamma_{12}(0) = 0$.

in Fig. 9-3(b) through 9-3(d). If the values of $|\gamma_{12}(0)|$ are calculated using Eq. (9-33) and plotted versus the separation, a pattern emerges like that shown in Fig. 9-4. The resulting pattern resembles very much the far-field pattern of a single aperture. In a fashion similar to that used in Chapter II, this pattern is related to the transform of source distribution. Thus this pattern will be shown to correspond to the spatial power spectrum of the source.

A method for making measurements of $|\gamma_{12}(\tau)|$ arises from the use of a Michelson interferometer (see Chapter VII) where time-delayed com-

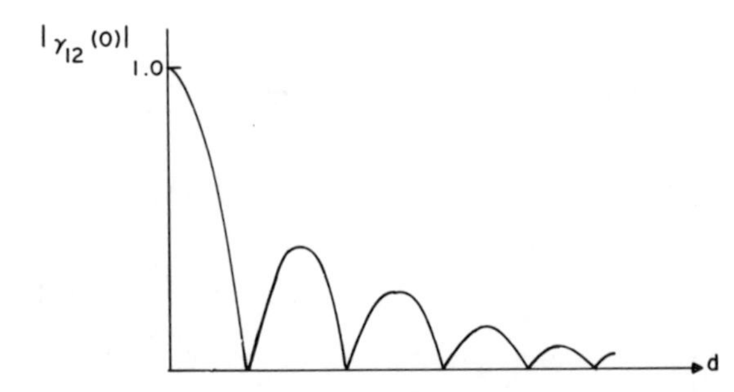

Fig. 9-4. Variation in the modulus of the degree of coherence versus sampling aperture separation.

parisons are made. To develop a clear understanding of this and of the processes and methods involved, a more concise formulation needs to be developed [59].

Propagation of Intensities

To understand more of how a partially coherent wave behaves, it is necessary to examine how the intensity distribution of one aperture propagates to other subsequent planes. To explore this notion, assume that a field V_p exists in a plane (x, y), of intensity $I_1(x, y)$. Assume that an output field V_o exists in a removed aperture located in a plane (x', y'), with intensity $I_2(x', y')$, and is excited by the field V_p. The governing equations are the wave equation and the Rayleigh–Sommerfeld equation depending on whether a differential or integral equation is used,

$$\nabla^2 V_i(x', y') + k^2 V_i(x', y') = 0 \tag{9-36}$$

or

$$V_o(x', y') = \frac{-j}{\lambda} p(z) \iint_{s_1} V_p(x, y) K(x' - x, y' - y)\, dx\, dy \tag{9-37}$$

In general, the term $(-j/\lambda)p(z)$ will be suppressed.

The input mutual coherence function $\Gamma_p(x, y)$ is given by

$$\Gamma_{p_{12}}(\tau) = \langle V_p(x_1, y_1, t) V_p^*(x_2, y_2, t + \tau)\rangle \tag{9-38}$$

which reduces to the self-intensity

$$I_p(x, y) = \langle V_p(x, y) V_p^*(x, y)\rangle \tag{9-39}$$

for $x_1 = x_2$, $y_1 = y_2$, and $t_1 = t_2$. The output self-intensity can be written in the same manner

$$I_o(x', y') = \langle V_o(x', y') V_o^*(x', y')\rangle \tag{9-40}$$

which, when Eq. (9-37) is applied, becomes†

$$I_o(x', y') = \Big\langle \iint V_p(x, y) K(x' - x, y' - y)\, dx\, dy \times \iint V_p^*(x'', y'') K^*(x' - x'', y' - y'')\, dx''\, dy'' \Big\rangle \tag{9-41}$$

† Reference to Fig. 1-2 is helpful for understanding the notation in each plane.

Now the amplitude impulse response of the system is represented by the kernel, $K(\ )$, and is time independent. Thus Eq. (9-41) can be changed to

$$I_0(x', y') = \iint_{s_1} \iint_{s_1} \langle V_p(x, y) V_p^*(x'', y'') \rangle K(x' - x, y' - y) \\ K(x' - x'', y' - y'') \, dx \, dy \, dx'' \, dy'' \tag{9-42}$$

In general, the bracketed term inside the integrals is the mutual intensity term.

This result enables an examination of the ideal cases of coherence and incoherence [101–103]. If incoherence is defined as the case when an average over all pairs of points is null except when $x = x''$ and $y = y''$, then the mutual intensity function can be written as

$$\langle V_p(x, y) V_p^*(x'', y'') \rangle = I_p(x, y) \, \delta(x - x'') \, \delta(y - y'') \tag{9-43}$$

Substituting Eq. (9-43) into Eq. (9-42) and integrating, we obtain a reduced form of $I_0(x', y')$:

$$I_0(x', y') = \iint_{s_1} I_p(x, y) K(x' - x, y' - y) K^*(x' - x, y' - y) \, dx \, dy \tag{9-44}$$

This equation can now be used to define an intensity impulse response for an incoherent system,

$$S(x' - x, y' - y) = K(x' - x, y' - y) K^*(x' - x, y' - y) \tag{9-45}$$

Thus $I_0(x', y')$ can be expressed in a simple convolutional form

$$I_0(x', y') = \iint_{s_1} I_p(x, y) S(x' - x, y' - y) \, dx \, dy \tag{9-46}$$

The similarity of this result with Eq. (9-37) is important since it allows linear analysis methods to be applied to intensities when the system is incoherent. Further, it gives an explicit integral equation form for describing the propagation of intensities.

If a coherent input is used, the intensity $I_p(x, y)$ is simply represented by the product of the input field with its conjugate

$$I_p(x, y) = U(x, y) U^*(x, y) \tag{9-47}$$

reducing Eq. (9-42) to

$$I_0(x', y') = \left| \iint_{s_1} U(x, y) K(x' - x, y' - y) \, dx \, dy \right|^2 \tag{9-48}$$

This result could have been obtained directly from the formulation represented by Eq. (9-37).

Thus intensities do propagate in a manner similar to fields, and the process can be represented by either of the integral expressions of Eq. (9-46) or Eq. (9-48). The case of the partially coherent wave is another matter. The propagation of partially coherent waves will first be examined from the wave equation perspective.

Propagation of the Mutual Coherence Function

The mutual coherence function $\Gamma_{12}(\tau)$ is composed of a time-averaged product of complex fields, Eq. (9-11). Each of these fields satisfy the wave equation,† Eq. (9-36). Thus the wave equation operator can be applied to each field variable. Explicitly this means applying the ∇^2 operator first over the (x_1, y_1) position variables and then over the variables (x_2, y_2). The spatial operator ∇_i^2 commutes with the time-averaging operator, giving

$$\nabla_1^2\Gamma_{12} = \langle \nabla_1^2 V(x_1, y_1, t_1) V^*(x_2, y_2, t_2) \rangle \tag{9-49a}$$

and

$$\nabla_2^2\Gamma_{12} = \langle V(x_1, y_1, t_1) \nabla_2^2 V(x_2, y_2, t_2) \rangle \tag{9-49b}$$

Using Eq. (9-36) in its time-differential form

$$\nabla_i^2 V(x_i, y_i, t_i) - \frac{1}{c^2} \frac{\partial^2 V(x_i, y_i, t_i)}{\partial t_i^2} = 0 \tag{9-50}$$

the spatial differentials can be replaced by

$$\nabla_1^2\Gamma_{12}(t_1) = \frac{1}{c^2} \frac{\partial^2}{\partial t_1^2} \langle V(x_1, y_1, t_1) V^*(x_2, y_2, t_2) \rangle \tag{9-51a}$$

and

$$\nabla_2^2\Gamma_{12}(t_2) = \frac{1}{c^2} \frac{\partial^2}{\partial t_2^2} \langle V(x_1, y_1, t_1) V^*(x_2, y_2, t_2) \rangle \tag{9-51b}$$

where the time average defined by Eq. (9-12) and the differentials over

† To be strictly correct, we should start with the fact that only the real part of V satisfies the wave equation since it is the real part of the complex or analytic field that is observable. However, since the real and imaginary parts of V are related by a Hilbert transform, then V also satisfies a wave equation like Eq. (9-50).

t_1 and t_2 have been commuted.† The operands on the right-hand side of Eqs. (9-51) are just the mutual coherence function Γ_{12}.

Now if the stationarity hypothesis is imposed, then

$$\partial^2/\partial t_1^2 = \partial^2/\partial t_2^2 = \partial^2/\partial\tau^2 \tag{9-52}$$

and Eqs. (9-51) reduce to

$$\nabla_s^2\Gamma_{12}(\tau) = \frac{1}{c^2}\frac{\partial^2\Gamma_{12}(\tau)}{\partial\tau^2} \qquad s = 1, 2 \tag{9-53}$$

where s is a subscript indicating which pair of spatial variables is being used in the differential. Thus, Eq. (9-53) represents two wave equations.

Thus, the mutual coherence function $\Gamma_{12}(\tau)$ satisfies a wave equation and propagates as did the intensity function. The distinction, however, is in the description of which four-dimensional space (x, y, z, t_i) each propagates. In the case of waves and intensities, the picture is classical. One can envision at a point in space a wave (intensity) train passing by; or conversely at a fixed instant of time t, wave (intensities) crests and valleys are distributed throughout space. However, in the case of the mutual coherence function, the four-dimensional space is physical space (x, y, z) plus relative time space τ. The relative time space is really related to differences in path length. Thus if one envisions a fixed point in τ, one is describing a locus of points lying on the contours‡ of hyperboloids, symmetric about the bisector plane of the sample points, that collapse into ellipsoids when the path difference exceeds the sample point separation in the input plane. This then gives an interesting equivalent picture where constant τ corresponds to motion over surfaces. However, since two of the variables describing the surfaces are also in the spatial differential variables, then constant τ points, for a given plane, are not simple crest and valley pictures, but rather complex loci in x and y. However, it is convenient to conceptualize a propagation process where a function does move over physical space as one varies the relative path difference to the observation

† This is valid since the correct time argument in $V(\ )$ is actually $t + t_1$ and $t + t_2$, where an origin in time space has been arbitrarily selected. The average in Eq. (9-12) is over the absolute time t.

‡ The equation describing these contours in the x', y', z space is

$$\frac{(x' - d/2)^2}{(\Delta l/2)^2} - \frac{y'^2 + z^2}{(d/2)^2 - (\Delta l/2)^2} = 1$$

where Δl is the path difference and d is the sample point separation.

point. That is, for a given delay there are points in the (x', y') plane where a particular mutual coherence can be found. One philosophical implication of this very abstract concept is that one cannot measure an incoherent source. For wherever the measurement is made, there is a locus of points for which $\Gamma_{ij}(\tau)$ is nonzero. A good example is that fringes can be formed with sunlight.

The Van Cittert–Zernike Theorem

In the last section, it was shown that the mutual coherence function satisfied a wave equation. Similarly, there must be a dual integral representation for $\Gamma_{12}(\tau)$ that parallels the dual representation for propagating fields and intensities. This integral equation representation can be derived by starting with Eq. (9-11) and using Eq. (9-37), giving

$$\Gamma_{o_{12}}(\tau) = \Big\langle (\,) \iint_s V_p(x, y, t) K(x_1 - x, y_1' - y)\, dx\, dy \times \iint V_p(x'', y'', t+\tau) K^*(x_2' - x'', y_2' - y'')\, dx''\, dy'' \Big\rangle \tag{9-54}$$

Using the properties of commutability of time and space operations, time independence of the kernels $K(\,)$, and the definition of $\Gamma_{p_{12}}$ from Eq. (9-11), we note that Eq. (9-54) reduces to

$$\Gamma_{o_{12}}(\tau) = (\,) \iint_s \iint_s \Gamma_{p_{12}}(\tau) K(x_1' - x, y_1' - y) \times K^*(x_2' - x'', y_2' - y'')\, dx\, dy\, dx''\, dy'' \tag{9-55}$$

This representation is similar to Eq. (9-44) derived for representing the propagation of intensities. It shows that $\Gamma_{p_{12}}(\tau)$ does indeed propagate; however, physical interpretation using this model is more difficult.

Equation (9-55) can be used, however, to relate $\Gamma_{o_{12}}(\tau)$ to a special case when a source of small extent is used, thus leading to negligible difference in the retardations r_1 and r_2. In this case $\Gamma_{p_{12}}(\tau)$ reduces to a product of the self-intensity of the source multiplied by a pair of delta functions as shown in Eq. (9-43). Thus Eq. (9-55) reduces to a single set of double integrals

$$\Gamma_{o_{12}}(\tau) = (\,) \iint_s I_p(x, y) K(x_1' - x, y_1' - y) K^*(x_2' - x, y_2' - y)\, dx\, dy \tag{9-56}$$

Note that in the case that (x_1', y_1') equals (x_2', y_2'), Eq. (9-56) reduces to Eq. (9-44), which was derived for propagating intensities.

The complex degree of coherence can be defined in terms of Eq. (9-56) if Eq. (9-14) is used with Eq. (9-44), giving

$$\gamma_{12}(\tau) = \frac{\iint_s I_p(x, y) K(x_1' - x, y_1' - y) K^*(x_2' - x, y_2' - y)\, dx\, dy}{\left[\iint_s I_p(x, y) \mid K_1 \mid^2 dx\, dy\right]^{1/2} \left[\iint_s I_p(x, y) \mid K_2 \mid^2 dx\, dy\right]^{1/2}} \tag{9-57}$$

where

$$\mid K_1 \mid^2 = K(x_1' - x, y_1' - y) K^*(x_1' - x, y_1' - y) \tag{9-58a}$$

and

$$\mid K_2 \mid^2 = K(x_2' - x, y_2' - y) K^*(x_2' - x, y_2' - y) \tag{9-58b}$$

Since this result is derived for a small source, a Huygens-type representation or point-source kernel can be used such that $\mid K_i \mid^2$ reduces to $1/R_i^2$, where R_i is the distance from the source point p to the sample points p_i. Applying this kernel K_i to the numerator, Eq. (9-57) reduces to

$$\gamma_{12}(\tau) = \frac{\iint_s I_p(x, y) \exp[ik(R_1 - R_2)]/R_1 R_2\, dx\, dy}{\left[\iint_s [I_p(x, y)]/R_1^2\, dx\, dy\right]^{1/2} \left[\iint_s [I_p(x, y)]/R_2^2\, dx\, dy\right]^{1/2}} \tag{9-59}$$

which is the more common form for the Van Cittert–Zernike theorem.

This theorem shows that the complex degree of coherence can be related to the source intensity distribution with appropriate normalization. The numerator is a diffraction-type integral similar to the earlier Rayleigh–Sommerfeld equation describing diffraction fields. The normalizing denominator is simply the source intensity reduced by the square of the amplitude spread function $1/R_i$. This result was first obtained by Van Cittert [95] (1934) and later by Zernike [94] (1938). It is, however, referred to by their combined names, Van Cittert–Zernike theorem [59].

Degree of Coherence and the Source Power Spectral Density

Further examination of Eq. (9-59) shows that if both R_1 and R_2 are very much greater than the aperture sizes, then the amplitude spread functions $1/R_i$ can be brought outside the integrals and canceled. This reduces Eq. (9-59) to

$$\gamma_{12}(\tau) = \frac{\iint_s I_p(x, y) \exp[ik(R_1 - R_2)]\, dx\, dy}{\iint_s I_p(x, y)\, dx\, dy} \tag{9-60}$$

If now the para-axial approximation for R_i is used, $R_1 - R_2$ becomes

$$R_1 - R_2 = \frac{(x_1'^2 - x_2'^2) + (y_1'^2 - y_2'^2)}{2z} - \frac{(x_1' - x_2')x + (y_1' - y_2')y}{z} \tag{9-61}$$

where the first term results in a pair of quadratic-phase terms that can be factored outside the integral. Using a polar representation r_i^2 for $(x_i^2 + y_i^2)$ and defining a phase function as

$$\phi_i = kr_i^2/2z \tag{9-62}$$

Eq. (9-60) reduces to

$$\gamma_{12}(\tau) = \frac{\exp[j(\phi_1 - \phi_2)] \iint_s I_p(x, y) \exp[ik(\xi x + \eta y)]\, dx\, dy}{\iint_s I_p(x, y)\, dx\, dy} \tag{9-63}$$

where

$$\xi = x_1' - x_2' \tag{9-64a}$$

and

$$\eta = y_1' - y_2' \tag{9-64b}$$

The numerator integral is in the form of a Fourier transform of the source power density function, where the transform variables are the difference coordinates defined in Eqs. (9-64). Thus, this representation shows that, except for a quadratic-phase term, the complex degree of coherence is the normalized power spectral density of the source density function $I_p(x, y)$. This then proves the result previously presented and shown in Fig. 9-4.

In this spectral representation, it is important to keep cognizance of the relative coordinates (ξ, η) which exist throughout the output plane (x', y'). Thus it is shown that this spectral measurement could only be made in a removed plane following a procedure similar to that shown in Fig. 9-3.

Imaging with Partially Coherent Light

The theory developed in the previous sections can now be applied to imaging with partially coherent light. Figure 9-5 shows a setup for studying propagating intensities of a finite diffraction-limited system, where the lens aperture function is defined by

$$L(\mathbf{x}'/a) = \begin{cases} 1 & |\mathbf{x}'| < a \\ 0 & |\mathbf{x}'| \geq a \end{cases} \tag{9-65}$$

and the lens has a focal length of f. The lens transmission will be represented by the classical quadratic-phase function, Eq. (3-10), where the constant phase terms will be dropped. It is assumed that a known $\Gamma_p(\mathbf{x}_1, \mathbf{x}_2, \tau)$ exists in the input plane (x, y) and that the output coherence function $\Gamma_0(\boldsymbol{\xi}_1, \boldsymbol{\xi}_2, \tau)$ is derived from this input function by the propagation process described in Eq. (9-55). Equation (9-55) has to be applied twice, first through the (x', y') plane and then to the output plane (ξ, η).

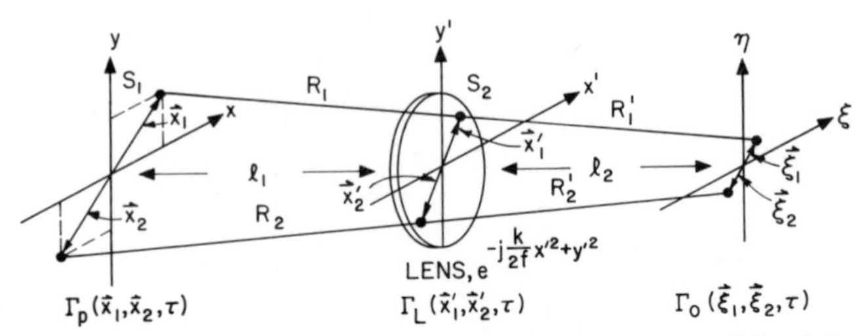

Fig. 9-5. A partially coherent imaging system having a lens of focal length f and finite extent $2a$.

Before writing out the explicit form of this representation, the propagator functions can be simplified for the case that l_i is very much greater than the transverse directions giving the para-axial approximation for

$$K(x' - x, y' - y)K^*(x' - x'', y' - y'') = \frac{\exp[jk(R_1 - R_2)]}{R_1 R_2} \tag{9-66}$$

as†

$$K(\,)K^*(\,) \cong \frac{\exp\{j(k/2l_1)[(\mathbf{x}_1' - \mathbf{x}_1)^2 - (\mathbf{x}_2' - \mathbf{x}_2)^2]\}}{l_1^2} \tag{9-67}$$

† A compact vector rotation is used for convenience, the explicit form is

$$[(\mathbf{x}_1' - \mathbf{x}_1)^2 - (\mathbf{x}_2' - \mathbf{x}_2)^2] = [x_1' - x_1)^2 + (y_1' - y_1)^2 - (x_2' - x_2)^2 - (y_2' - y_2)^2]$$

where Eq. (9-55) now becomes

$$\Gamma_L(\mathbf{x}_1', \mathbf{x}_2', \tau) = \frac{(\ )}{l_1^2} \iint_{s_1} \iint_{s_1} \Gamma_p(\mathbf{x}_1, \mathbf{x}_2, \tau) \times \exp\left\{ j \frac{k}{2l_1} [(\mathbf{x}_1' - \mathbf{x}_1)^2 - (\mathbf{x}_2' - \mathbf{x}_2)^2] \right\} dx_1\, dy_1\, dx_2\, dy_2 \tag{9-68}$$

Applying this result a second time and accounting for the finite lens, we note that the output mutual coherence function becomes

$$\begin{aligned}\Gamma_o(\boldsymbol{\xi}_1, \boldsymbol{\xi}_2, \tau) = \frac{(\ )}{l_1^2 l_2^2} &\iint_{s_1} \iint_{s_1} \iint_{s_2} \iint_{s_2} \Gamma_p(\mathbf{x}_1, \mathbf{x}_2, \tau) \\ &\times \exp\left\{ j \frac{k}{2l_1} [(\mathbf{x}_1' - \mathbf{x}_1)^2 - (\mathbf{x}_2' - \mathbf{x}_2)^2] \right\} \\ &\times \exp\left[-j \frac{k}{2f} (x_1'^2 + y_1'^2) \right] \exp\left[j \frac{k}{2f} (x_2'^2 + y_2'^2) \right] \\ &\times L\left(\frac{\mathbf{x}_1'}{a}\right) L\left(\frac{\mathbf{x}_2'}{a}\right) \exp\left\{ j \frac{k}{2l_2} [(\boldsymbol{\xi}_1 - \mathbf{x}_1')^2 - (\boldsymbol{\xi}_2 - \mathbf{x}_2')^2] \right. \\ &\left. \times dx_1\, dy_1\, dx_2\, dy_2\, dx_1'\, dy_1'\, dx_2'\, dy_2' \right\}\end{aligned} \tag{9-69}$$

where the lens aperture and transmission functions have been applied twice corresponding to the two field-variable operations represented by the mutual coherence function. Equation (9-69) can be factored and rearranged to make the imaging constraints more obvious as follows:

$$\begin{aligned}\Gamma_o(\boldsymbol{\xi}_1, \boldsymbol{\xi}_2, \tau) = \frac{j(k/2l_2)(\boldsymbol{\xi}_1^2 - \boldsymbol{\xi}_2^2)}{l_1^2 l_2^2} &\iint_{s_1} \iint_{s_1} \iint_{s_2} \iint_{s_2} dx_1\, dy_1\, dx_2\, dy_2 \\ &\times dx_1'\, dy_1'\, dx_2'\, dy_2' \Gamma_p(\mathbf{x}_1, \mathbf{x}_2, \tau) L\left(\frac{\mathbf{x}_1'}{a}\right) L^*\left(\frac{\mathbf{x}_2'}{a}\right) \\ &\times \exp\left[\frac{jk\mathbf{x}_1'^2}{2} \left(\frac{1}{l_1} + \frac{1}{l_2} - \frac{1}{f} \right) + j \frac{k\mathbf{x}_1^2}{2l_1} - jk\mathbf{x}_1' \right. \\ &\left. \times \left(\frac{\mathbf{x}_1}{l_1} + \frac{\boldsymbol{\xi}_1}{l_1} \right) \right] \exp\left[\frac{-jk\mathbf{x}_2'^2}{2} \left(\frac{1}{l_1} + \frac{1}{l_2} - \frac{1}{f} \right) \right. \\ &\left. - j \frac{k\mathbf{x}_2^2}{2l_1} + jk\mathbf{x}_2' \cdot \left(\frac{\mathbf{x}_2}{l_1} + \frac{\boldsymbol{\xi}_2}{l_2} \right) \right]\end{aligned} \tag{9-70}$$

If the image condition defined by Eq. (3-22) is applied,

$$1/l_1 + 1/l_2 = 1/f \tag{9-71}$$

and Eq. (9-70) is then rearranged, the output mutual coherence function reduces to the product of three terms:

$$\begin{aligned} \Gamma_o(\xi_1, \xi_2, \tau) = {} & \left\{ \frac{\exp[j(k/2l_2)(\xi_1^2 - \xi_2^2)]}{l_1^2 l_2^2} \right\} \iint_{s_1} \iint_{s_2} dx_1\, dy_1\, dx_2\, dy_2 \\ & \times \Gamma_p(\mathbf{x}_1, \mathbf{x}_2, \tau) \exp\left[\frac{jk}{2l_1} (\mathbf{x}_1^2 - \mathbf{x}_2^2) \right] \\ & \times \iint_{s_2} dx_1'\, dy_1'\, L\left(\frac{\mathbf{x}_1'}{a} \right) \exp\left[-jk\mathbf{x}_1 \cdot \left(\frac{\mathbf{x}_1}{l_1} + \frac{\xi_1}{l_2} \right) \right] \\ & \times \iint_{s_2} dx_2'\, dy_2'\, L^*\left(\frac{x_2'}{a} \right) \exp\left[jk\mathbf{x}_2' \cdot \left(\frac{\mathbf{x}_2}{l_1} + \frac{\xi_2}{l_2} \right) \right] \end{aligned} \tag{9-72}$$

where the last two are complex conjugates. The last two terms are also Fourier transforms of the lens aperture functions, which by definition are the amplitude impulse response of the imaging system $K(\ )K^*(\ )$, where the argument of $K(\ )$ is $(\xi_1/l_2 + \mathbf{x}_i/l_1)$. Thus the image equation for the mutual coherence function can be written

$$\begin{aligned} \Gamma_o(\xi_1, \xi_2, \tau) = {} & \left\{ \frac{\exp[j(k/2l_2)(\xi_1^2 - \xi_2^2)]}{l_1^2 l_2^2} \right\} \iint_{s_1} \iint_{s_2} dx_1\, dy_1\, dx_2\, dy_2 \\ & \times \Gamma_p(\mathbf{x}_1, \mathbf{x}_2, \tau) \exp\left[\frac{jk}{2l_1} (\mathbf{x}_1^2 - \mathbf{x}_2^2) \right] \\ & \times K\left(\frac{\xi_1}{l_2} + \frac{\mathbf{x}_1}{l_1} \right) K^*\left(\frac{\xi_2}{l_2} + \frac{\mathbf{x}_2}{l_1} \right) \end{aligned} \tag{9-73}$$

Thus except for the quadratic-phase factors, the output mutual coherence function is expressed as a function of the input coherence function where the major factor in the integrand is the product of the amplitude impulse responses. These terms indicate that the output function is inverted and magnified by the ratio l_2/l_1.

For one-dimensional systems, these impulse response functions for finite apertures are sinc functions. For two-dimensional circular apertures, these are jinc functions† [104], which under large-argument approximations reduce

† See Eq. (2-29).

to delta functions. Thus if the coherence interval at the input object is large compared to the distances resolved by the imaging system, then the impulse responses can be represented by delta functions, giving

$$\Gamma_o(\boldsymbol{\xi}_1, \boldsymbol{\xi}_2, \tau) = \{\ \} \iint_{s_1} \iint_{s_2} dx_1\, dy_1\, dx_2\, dy_2\, \Gamma_p(\mathbf{x}_1, \mathbf{x}_2, \tau) \times \exp\left[\frac{jk}{2l_1}(\mathbf{x}_1{}^2 - \mathbf{x}_2{}^2)\right] \delta\left(\frac{\boldsymbol{\xi}_1}{l_2} + \frac{\mathbf{x}_1}{l_1}\right) \delta\left(\frac{\boldsymbol{\xi}_2}{l_2} + \frac{\mathbf{x}_2}{l_1}\right) \tag{9-74}$$

which, upon integration, yields

$$\Gamma_o(\boldsymbol{\xi}_1, \boldsymbol{\xi}_2, \tau) = (\ ) \exp\left[jk\left(1 - \frac{z_1}{z_2}\right)\left(\frac{\boldsymbol{\xi}_1{}^2 - \boldsymbol{\xi}_2{}^2}{2l_2}\right)\right] \times \Gamma_p\left(\frac{-l_1}{l_2}\boldsymbol{\xi}_1, \frac{-l_1}{l_2}\boldsymbol{\xi}_2\right) \tag{9-75}$$

This result is explicitly like the imaging result obtained in Chapter II for coherent systems. The image function is inverted and scaled by the magnification ratio l_2/l_1. For the case of unit magnification, the quadratic-phase term disappears, and a simple inverted image function remains:

$$\Gamma_o(\boldsymbol{\xi}_1, \boldsymbol{\xi}_2, \tau) = \Gamma_p(-\boldsymbol{\xi}_1, -\boldsymbol{\xi}_2, \tau) \tag{9-76}$$

This is not the coherent limit, but rather a case where fine detail can be resolved.

For the case where coherence intervals cannot be resolved by the imaging system, the input coherence function can be replaced by the incoherent expression, or self-intensity of the input system, reducing Eq. (9-73) to

$$\Gamma_o(\boldsymbol{\xi}_1, \boldsymbol{\xi}_2, \tau) = \{\ \} \iint \iint dx_1\, dy_1\, dx_2\, dy_2\, I_p(\mathbf{x}_1)\, \delta(x_1 - x_2)\, \delta(y_1 - y_2) \times K\left(\frac{\boldsymbol{\xi}_1}{l_2} + \frac{\mathbf{x}_1}{l_1}\right) K^*\left(\frac{\boldsymbol{\xi}_2}{l_2} + \frac{\mathbf{x}_2}{l_1}\right) \tag{9-77}$$

which gives

$$\Gamma_o(\boldsymbol{\xi}_1, \boldsymbol{\xi}_2, \tau) = \{\ \} \iint_{s_1} dx\, dy\, I_p(x, y) K\left(\frac{\boldsymbol{\xi}_1}{l_2} + \frac{\mathbf{x}_1}{l_1}\right) K^*\left(\frac{\boldsymbol{\xi}_2}{l_2} + \frac{\mathbf{x}_2}{l_1}\right) \tag{9-78}$$

The intensity of the output distribution can be recovered by setting $\boldsymbol{\xi}_1$

equal to $\boldsymbol{\xi}_2$, yielding

$$I_o(\xi, \eta) = \iint_{s_1} dx\, dy\, I_p(x, y) \left| K\left(\frac{\boldsymbol{\xi}}{l_2} + \frac{\mathbf{x}}{l_1}\right)\right|^2 \tag{9-79}$$

which is consistent with the incoherent limit expressed in Eq. (9-56) and the propagation of intensities result, Eq. (9-46).

Equation (9-73) can be examined from another point of view, which does not depend on the two approximations described above. To do so, the image intensity distribution is obtained from Eq. (9-31) by setting $\boldsymbol{\xi}_1$ equal to $\boldsymbol{\xi}_2$, yielding

$$I_o(\boldsymbol{\xi}) = (\,) \iint \iint dx_1\, dy_1\, dx_2\, dy_2\, \Gamma_p(\mathbf{x}_1, \mathbf{x}_2, \tau) \exp\left[\frac{jk}{2l_1}(\mathbf{x}_1{}^2 - \mathbf{x}_2{}^2)\right]$$

$$\times K\left(\frac{\boldsymbol{\xi}}{l_2} + \frac{\mathbf{x}_1}{l_1}\right) K^*\left(\frac{\boldsymbol{\xi}}{l_2} + \frac{\mathbf{x}_2}{l_1}\right) \tag{9-80}$$

Now the coherent limit can be explored by substituting Eq. (9-47), factoring and rearranging the complex conjugate terms to obtain

$$I_o(\boldsymbol{\xi}) = (\,) \left| \iint_{s_1} dx\, dy\, U(\mathbf{x}) \exp\left(\frac{jk\mathbf{x}^2}{2l_1}\right) K\left(\frac{\boldsymbol{\xi}}{l_2} + \frac{\mathbf{x}}{l_1}\right)\right|^2 \tag{9-81}$$

which, for a one-dimensional case, reduces to

$$I_o(\xi) = (\,) \left| \int dx\, U(x) \exp\left[\frac{jk\mathbf{x}^2}{2l_1}\right] a \operatorname{sinc}\left[ka\left(\frac{\xi}{l_2} + \frac{x}{l_1}\right)\right]\right|^2 \tag{9-82}$$

If ka is large, then the sinc function reduces to a delta function, and the simple result is obtained that

$$I_o(\xi) = \left| U\left(\frac{l_1}{l_2}\, \xi\right)\right|^2 \tag{9-83}$$

which is the coherent limit that should be contrasted with Eq. (9-76).

The incoherent limit can be obtained by applying Eq. (9-43) to Eq. (9-80). The result is the same as that derived in Eq. (9-79). This consistency in form is important when attempting to understand the implications of partially coherent imaging. In addition, Eq. (9-78) points out again that even though an input object is incoherent, the output is not incoherent. This result is particularly important in cascaded optical systems that are partially coherent. Thus strictly speaking, incoherent transfer functions cannot be multiplied together [105–107].

Fourier Transforms with Partially Coherent Light

A Fourier-transforming system can be considered [105] by following the development in Chapters II and III and setting l_1 and l_2 equal to f in the system shown in Fig. 9-5. This adjustment in axial separation removes the quadratic-phase factors and simplifies the basic equation, Eq. (9-69). Further, it will be assumed that the input mutual coherence function is associated with transilluminated input objects that are represented by complex amplitude transmission masks $t(\mathbf{x})$. Thus Eq. (9-69) can be simplified and rearranged to[†]

$$\Gamma_o(\boldsymbol{\xi}_1, \boldsymbol{\xi}_2) = \frac{1}{f^4} \iint_{s_1} \iint_{s_2} \Gamma_p(\mathbf{x}_1 - \mathbf{x}_2) t(\mathbf{x}_1) t^*(\mathbf{x}_2) \times K(\boldsymbol{\xi}_1 - \mathbf{x}_1) K^*(\boldsymbol{\xi}_2 - \mathbf{x}_2)\, dx_1\, dy_1\, dx_2\, dy_2 \tag{9-84}$$

where spatial stationarity and invariance have been assumed. This is particularly important when examining the question of transfer function representation. Equation (9-84) reduces to the output intensity distribution when $\boldsymbol{\xi}_1$ equals $\boldsymbol{\xi}_2$. This then allows an examination of the transfer function concept which is so important in linear system analysis.

Transfer function concepts follow quite naturally from spectrum concepts. Thus, an input object spectrum can be defined as

$$\hat{I}_p(\boldsymbol{\mu}) = \iint dx\, dy\, I_p(\mathbf{x}) \exp(-j2\pi\boldsymbol{\mu} \cdot \mathbf{x}) \tag{9-85}$$

and an output spectrum as

$$\hat{I}_o(\boldsymbol{\sigma}) = \iint d\xi\, d\eta\, I_o(\boldsymbol{\xi}) \exp(-j2\pi\boldsymbol{\sigma} \cdot \boldsymbol{\xi}) \tag{9-86}$$

Applying Eq. (9-86) to Eq. (9-84) leads to

$$\hat{I}_o(\boldsymbol{\sigma}) = \iint_{s_1} \iint_{s_1} \iint_{s_2} \Gamma_p(\mathbf{x}_1 - \mathbf{x}_2) t(\mathbf{x}_1) t^*(\mathbf{x}_2) K(\boldsymbol{\xi} - \mathbf{x}_1) K^*(\boldsymbol{\xi} - \mathbf{x}_2) \times \exp(-j2\pi\boldsymbol{\sigma} \cdot \boldsymbol{\xi})\, d\xi\, d\eta\, dx_1\, dy_1\, dx_2\, dy_2 \tag{9-87}$$

which results in a form of Fourier transform on the product of the amplitude impulse responses. This transform of products can also be expressed as a

[†] The τ in the argument of $\Gamma(\,)$ has been suppressed for convenience, since it is implicit.

convolution of individual transforms when the appropriate transformations and conjugate variables are accounted for. Thus, the Fourier-transform term can be expressed as

$$\iint_{s_2} d\xi\, d\eta\, K(\boldsymbol{\xi} - \mathbf{x}_1)K^*(\boldsymbol{\xi} - \mathbf{x}_2) \exp(-j2\pi\boldsymbol{\sigma} \cdot \boldsymbol{\xi})$$
$$= \exp(-j2\pi\boldsymbol{\sigma} \cdot \mathbf{x}_1) \iint d\mathbf{p}\, \hat{K}(\boldsymbol{\sigma} - \mathbf{p})\hat{K}^*(\mathbf{p}) \exp[j2\pi\mathbf{p} \cdot (\mathbf{x}_1 - \mathbf{x}_2)] \tag{9-88}$$

where

$$\hat{K}(\boldsymbol{\sigma} - \mathbf{p}) = \exp[j2\pi(\boldsymbol{\sigma} - \mathbf{p}) \cdot \boldsymbol{\xi}_1] \iint d\mathbf{x}\, K(\mathbf{x} - \boldsymbol{\xi}_1) \times \exp[-j2\pi(\boldsymbol{\sigma} - \mathbf{p}) \cdot \mathbf{x}] \tag{9-89a}$$

and

$$\hat{K}(\mathbf{p}) = \exp(j2\pi\mathbf{p} \cdot \boldsymbol{\xi}_2) \iint d\mathbf{x}\, K(\mathbf{x} - \boldsymbol{\xi}_2) \exp(-j2\pi\mathbf{p} \cdot \mathbf{x}) \tag{9-89b}$$

Thus Eq. (9-87) becomes

$$\hat{I}_o(\boldsymbol{\sigma}) = \iint_{s_1} \iint \hat{K}(\boldsymbol{\sigma} - \mathbf{p})\hat{K}^*(\mathbf{p})t^*(\mathbf{x}_2)$$
$$\times \left[\iint_{s_1} \Gamma_p(\mathbf{x}_1 - \mathbf{x}_2)t(\mathbf{x}_1) \exp[-j2\pi(\boldsymbol{\sigma} - \mathbf{p}) \cdot \mathbf{x}_1\, dx_1]\right]$$
$$\times \exp(-j2\pi\mathbf{p} \cdot \mathbf{x}_2)\, d\mathbf{p}\, d\mathbf{x}_2 \tag{9-90}$$

The bracketed term, [ ], can also be written in terms of a Fourier transform by writing

$$\iint \Gamma_p(\mathbf{x}_1 - \mathbf{x}_2)t(\mathbf{x}_1) \exp[-j2\pi(\boldsymbol{\sigma} - \mathbf{p}) \cdot \mathbf{x}_1\, d\mathbf{x}_1]$$
$$= \exp[-j2\pi(\boldsymbol{\sigma} - \mathbf{p}) \cdot \mathbf{x}_2] \iint \hat{\Gamma}_p(\boldsymbol{\sigma} - \mathbf{p} - \mathbf{v})\hat{t}(\mathbf{v}) \exp(j2\pi\mathbf{v} \cdot \mathbf{x}_2)\, d\mathbf{v} \tag{9-91}$$

which, when substituted into Eq. (9-90), yields

$$\hat{I}_o(\boldsymbol{\sigma}) = \iint \iint \hat{\Gamma}_p(\boldsymbol{\sigma} - \mathbf{p} - \mathbf{v})\hat{K}(\boldsymbol{\sigma} - \mathbf{p})\hat{K}^*(\mathbf{p})\hat{t}(\mathbf{v})$$
$$\times \left[\iint t^*(\mathbf{x}_2) \exp[-j2\pi\mathbf{x}_2 \cdot (\boldsymbol{\sigma} - \mathbf{v})\, d\mathbf{x}_2]\right] d\mathbf{p}\, d\mathbf{v} \tag{9-92}$$

Here again another bracketed term is created which enables the expression of $t^*(\mathbf{x}_2)$ as a Fourier transform. Performing the indicated transform and substituting the result produce the desired spectral representation of the output intensity

$$\hat{I}_0(\boldsymbol{\sigma}) = \iint \hat{t}(\mathbf{v})\hat{t}^*(\boldsymbol{\sigma}-\mathbf{v})\left[\iint \hat{\Gamma}_p(\boldsymbol{\sigma}-\mathbf{p}-\mathbf{v})\hat{K}(\boldsymbol{\sigma}-\mathbf{p})\hat{K}^*(\mathbf{p})\,d\mathbf{p}\right]d\mathbf{v} \qquad (9\text{-}93)$$

The inner bracketed integral is the characteristic of the partially coherent imaging system. Though like a convolution, it is not exactly in the proper form since three terms are shifted differently over three different vector spaces. The total bracketed term is a function of $\boldsymbol{\sigma}$ and $\mathbf{v}$, which, when included in the outer integral with the transmission functions, gives a quasi-convolutional form for output intensity. For a completely linear analysis form, one desires an output spectrum to be simply represented by the product of an object spectrum and transfer function. In this case, a multidimensional convolutional form is obtained.

Thus one cannot, in general, consider partially coherent systems as linear systems. The concept of "generalized transfer function" has been applied [59] to Eq. (9-93). This is a misleading notion since the transfer function concept was not established. It is important to note that no matter whether one is in the spatial or spectral domain, some convolutional-like integral remains [103]. Thus the partially coherent system is most appropriately labeled as a nonlinear system.

These results are, however, useful for representing what happens in a partially coherent system [105]. Of particular interest is how the classical resolution criteria have to be changed to compare coherent, partially coherent, and incoherent objects [108]. Beran and Parrent [59] and others have compared the resolution criteria and resolvability of two-point objects. They showed that the incoherent two-point resolution of a given optical system over a coherent system is better by a factor of 1.56 and that an almost linear relationship exists for this change of resolution between the incoherent and coherent range of $|\gamma|$. The major point is that reliance on classical resolution criteria can be misleading since not enough information is available for complete parameter measurement. An example would be in slit width and edge measurements [105]. Thus, much remains to be done in formulating and testing a more complete theory for describing and using partially coherent optical systems [109].

Hanbury-Brown and Twiss Experiment

So far in this chapter the coherence function $\Gamma_{12}(\tau)$ was the primary focus of the presentation and theoretical development. This is primarily because of its theoretical simplicity and ease with which it can be measured. The correlation represented by $\Gamma_{12}(\tau)$ is, however, a second-order moment over the field variables, which leads to the question of examining higher-order moments or correlations. Hanbury Brown and Twiss [110–119] and others showed in a series of experiments (1952–1956) that the correlation over intensities

$$R_{ij}(\mathbf{x}_1, \mathbf{x}_2) = \langle I_i(\mathbf{x}_1, t) I_j(\mathbf{x}_2, t)\rangle \tag{9-94}$$

where $I_i(\mathbf{x}_i, t)$ and $I_j(\mathbf{x}_2, t)$ are instantaneous values, could be used to infer information about the complex degree of coherence $\gamma_{12}(\tau)$ from the fluctuating part of the intensity correlation.

The problem is difficult, however, because it is not easy to obtain average and wideband fluctuating components of intensity. Note that in the case of $\Gamma_{12}(\tau)$, average values of $V(\mathbf{x}_i, t_i)$ were easy to obtain by the intensity detection process. No such equivalent process exists for intensities.

The correlation of intensities is a fourth-order moment, which in general is not related to the second-order moments; however, for gaussian statistics there is a simple relationship between the two correlations [120].

It can be shown [121] that the intensity correlation, Eq. (9-94), for gaussian statistics can be related to the degree of coherence by

$$R_{ij}(\mathbf{x}_1, \mathbf{x}_2, \tau) = \bar{I}_1 \bar{I}_2 [1 + |\gamma_{12}(\tau)|^2] \tag{9-95}$$

where $\bar{I}_1$ and $\bar{I}_2$ are the mean values of I_1 and I_2.

This can be further simplified by examining the correlation of the fluctuations, which are defined by subtracting from the measured intensities the average values, leaving

$$\langle \Delta I_1(t+\tau)\, \Delta I_2(t)\rangle = R_{ij}(\mathbf{x}_1, \mathbf{x}_2, \tau) - \bar{I}_1 \bar{I}_2 = \bar{I}_1 \bar{I}_2 |\gamma_{12}(\tau)|^2 \tag{9-96}$$

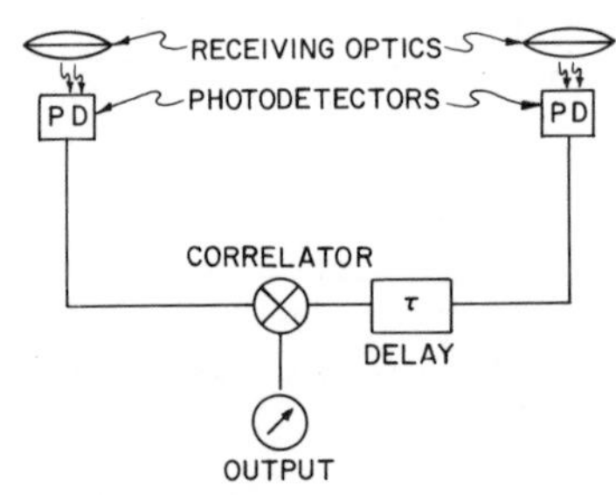

Fig. 9-6. A simplified intensity interferometer of the type used by Hanbury-Brown and Twiss. A correlation over intensities.

which is the real contribution of the Hanbury Brown and Twiss experiments from a wave point of view.

The experimental setup is shown in Fig. 9-6 and basically consists of a system in which intensity detection occurs prior to correlation. One of the intensity signals is delayed, and then the two intensities are correlated, producing a signal proportional to Eq. (9-95). These experiments enabled another approach to the study of partially coherent waves through the use of intensity interferometry.

Summary

In this chapter an effort has been made to develop a method for describing waves that are neither coherent nor incoherent, but rather have coherence properties that can vary over a considerable range. The measure of these variations was related to the ability of these waves to form fringes when interacting with sampling systems. Concepts of fringe visibility were related to the complex degree of coherence function.

Propagation properties of waves, intensities, and mutual coherence functions were explored. They were all shown to satisfy wave-type equations with some modification to the physical interpretation when describing mutual coherence function propagation properties.

Imaging and Fourier-transforming properties at partially coherent systems were examined and shown to reduce under special conditions to known properties of intensity descriptions. The Fourier-transforming investigations led to the realization that the linear transform methods cannot generally be applied to partially coherent systems. The fact that no general classical transfer function existed for a partially coherent system has reduced some of the flexibility offered by linear analysis methods. This constraint, however, is only a general one, and much has been done to study special transfer functions [101] for partially coherent systems, such as microdensitometers [102, 122] and stellar viewing systems [123].

Finally, the Hanbury-Brown and Twiss intensity interferometry experiments were discussed. The correlation between the fourth-order moments measured through intensity fluctuation correlations and the second-order degree of coherence function was noted. There is much to be done in intensity interferometry since it holds considerable promise in determining partially coherent source characteristics [60, 61].

Problems

1. Consider the cascaded optical system shown in Fig. 9-7, where l_1, l_2, $l_3 \gg d$; $c \ll d$; and $c \gg \lambda$. Calculate the field at the plane (x'', y''). Plot the variation of this field for various values of d; $c \leq d \leq 10a$ in the one dimension, y''. (*Hint*: Think of the slits as a sampling system.)

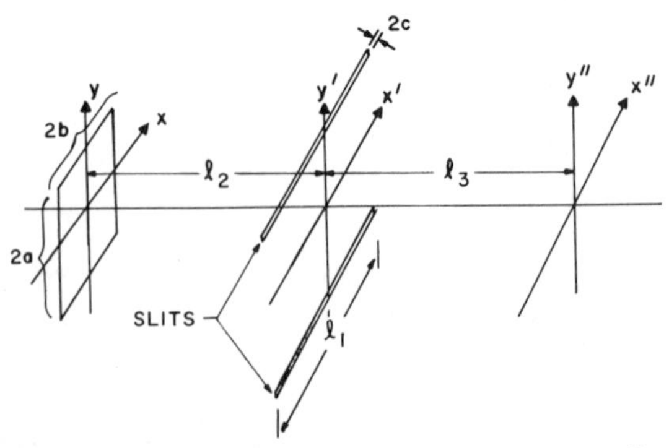

Fig. 9-7. An optical system for sampling the input aperture distribution.

2. Show that for a circular source aperture of radius p, that

$$\gamma_{12} \sim \left[\frac{2J_1(kpr/z)}{kpr/z}\right] \exp[i(\phi_1 - \phi_2)]$$

where r is the difference coordinate in the output plane. From this expression find the position r for the first null, in terms of p, z, and $\bar{\lambda}$.

3. Show that for two small apertures separated by a distance d in the (x, y) plane, the contours of constant delay τ in the (x', y', z) plane are described by the equation

$$\frac{(x' - d/2)^2}{(\Delta l/2)^2} - \frac{y'^2 + z^2}{(d/2)^2 - (\Delta l/2)^2} = 1$$

4. Derive Eq. (9-22) using the results of Eq. (9-21).

5. Derive Eq. (9-33) using Eq. (9-31) and Eq. (9-32).

6. Show in detail each of the Fourier transforms leading to Eq. (9-93).

7. Derive Eq. (9-88) by showing in detail the arguments leading to Eqs. (9-89).

Chapter X

SCATTERING

Introduction

An introduction to applied optics would be incomplete without some mention of scattering phenomena and processes. This is particularly true since so many observable optical phenomena in the everyday world are simple scattering processes. Examples are the blue sky, the white clouds, the red sunsets, the rainbow, the twinkling stars, and many more. These common scattering processes are some of the oldest studied phenomena (Da Vinci, 1500). There are even indications that scholars studied these phenomena in Biblical times; however, it is not until post-Renaissance that scientifically recorded work begins. In the middle 1800s, some of the most succinct observations were made. Tyndall (1869), Govi (1860), and Brücke (1853) observed that small particles scatter blue light. In fact it was Tyndall's work that helped Lord Rayleigh [124] (1899) in his simple explanation of the blue sky. From these early observations it was also possible to explain the observed polarization of the sky light.

When the particles are not small, the scattering changes and does not depend so strongly on wavelength. The explanation was basically provided by Mie [125] and showed that the scattering for larger particles depended on a ratio of particle size to wavelength, with an asymptotic behavior for large ratio values. These results then showed that the inverse fourth-power dependence on wavelength could also occur at other wavelengths for some particles.

An interesting result of these scattering theories is that one can use these phenomena to determine the nature of the scatterer. This inverse problem has been the focus of much research[†] since the turn of the century. It has particular significance today in questions of remote probing and studies of biological systems [40].

Probing of the atmosphere and attempts to propagate laser beams in an unperturbed fashion through the atmosphere have encountered problems with scattering due to the turbulence of the atmosphere. Even though the atmosphere as a dielectric is a weak scatterer in the average sense (less than one part in a hundred change), the total path length effect can significantly degrade a laser beam. This degradation can be severe to the point that an image cannot be propagated more than a few meters without considerable or even complete corruption. The study of how these wave parameter perturbations depend on the turbulence and medium properties has attracted considerable attention in recent years [126].

With this brief introduction to scattering, one can proceed to more advanced work [4] in scattering theory and its application. In particular, this chapter ends with brief consideration of the scattering formulations in terms of intensities and radiative transfer [127]. This approach leads to an expanded integral equation formulation [128–130].

Blue Sky

It is easiest to begin with an explanation of the blue sky. It is a phenomenon with which everyone is familiar and can first be argued from a dimensional analysis point of view. This argument was first put forth by Lord Rayleigh [124] and depends on the existence of small, compared to wavelength, scatterers of dilute concentration. A more detailed argument will be presented in a later section.

The scattered intensity I is assumed to be proportional to the incident intensity I_0 by a separable scattering function $f(\,)$. This scattering function depends on several factors. The primary factors are the volume of the scatterer V, the refractive indices of the scatterer η_2 and the medium η_1, the distance to the observer r, and the wavelength λ. Formally, this is written

$$I = f(V, \eta_1, \eta_2, r, \lambda) I_0 \tag{10-1}$$

The dependence on r is the easiest to establish, since a small pointlike scatterer is assumed. As the energy propagates out from this small source point, the surface area of the propagating sphere is expanding in proportion

[†] See book by M. Kerker [4], p. 311.

to r^2. Conservation of energy requires that the intensity must decrease as r^{-2}. This argument can be further substantiated by noting that if the scattering is due to induced dipoles,† the scattered field has a $1/r$ dependence, which leads to the same r^{-2} dependence. The dependence on volume is also related to the assumed dipole scattering. The scattered field is proportional to the total dipole moment, which is, in turn, directly proportional to volume. Thus, the volume dependence for the intensity is V^2. Because η_1 and η_2 are nondimensional, the dependence on η_1 and η_2 is also nondimensional. For the moment then, the dependence on η_1 and η_2 will be described by a scattering function $f_1(\eta_1, \eta_2)$. Since the total scatter function $f(\ )$ is nondimensional, the dependence on r, V, and λ must be nondimensional. Also, since the dependence on V and r goes as V^2/r^2 or a fourth-power dependence on length, the dependence on the remaining parameter λ must be λ^{-4}. Thus the scattered intensity can be written

$$I = f_1(\eta_1, \eta_2)(V^2/r^2\lambda^4)I_0 \tag{10-2}$$

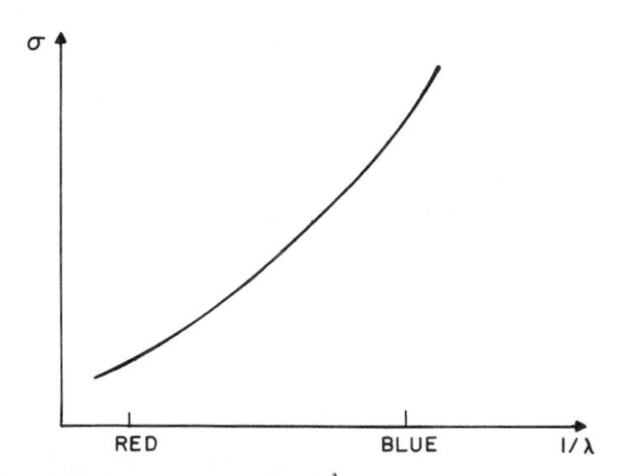

Fig. 10-1. Rayleigh region scattering efficiency σ versus inverse wavelength $1/\lambda$.

An alternate way of expressing this is in terms of the scattering efficiency σ, which is a normalized expression,

$$\sigma = I/I_0 = f_1(\eta_1, \eta_2)(V^2/r^2\lambda^4) \tag{10-3}$$

A rough sketch of this inverse fourth-power wavelength dependence is shown in Fig. 10-1. This figure illustrates very graphically why the blue light is more strongly scattered out of the direct illumination. Thus, when

† This follows closely the assumption of particle size being much smaller than the wavelength. In this case, the particle is then influenced by a uniform field, and it is analogous to the well-known electrostatic dipole formation.

one views the sky under nondirect illumination, the blue light is heavily emphasized when the scatterers are primarily the small aerosols and air molecules.

Red Sunset

The explanation of the red sunset follows quite naturally from the blue sky arguments. If the blue light is strongly scattered out of the direct forward path of illumination, then the direct observation of the sunset is dominated by the red end of the visible spectrum. This is even more apparent due to the long path length of the rays in the atmosphere at the low horizon angles. One problem is that the atmosphere is not always a clear dilute suspension of small scatterers; thus the red effect is not solely dependent on this process.

There is, however, a simple verification of red sunset being dominated by Rayleigh scattering. If one is standing, facing the west, with a red sunset on a reasonably clear evening, turning to observe either the north or south sky reveals a beautiful blue sky. This blueness arises from the strong scattering of the shorter wavelengths out of the direct parallel illumination path in a direction perpendicular to the observer.

Polarization of Skylight

Lord Rayleigh developed a very simple argument to explain polarization using much of the work related to the blue sky explanation. It follows that, if one again assumes a dilute suspension of small particles illuminated with a plane wave, that a simple excited dipole system can be assumed. In Fig. 10-2, the system is shown with the incident plane wave along the z axis. This plane wave excites an instantaneous dipole in the xy plane that reradiates the secondary emission that gives rise to the scattered wave. Since the incident wave can have arbitrary or random polarization, the excitation needs to be decomposed into two linear components. These components can arbitrarily be along η and β. If the observer is located at a position such that the wave vector to the observer makes an angle θ with respect to z axis, then the plane wave received by the observer will contain the transverse components η' and $\beta' \cos\theta$. The longitudinal components with respect to the observer will not contribute since the far-field electromagnetic plane wave has no longitudinal components. As θ approaches 90°, then only the

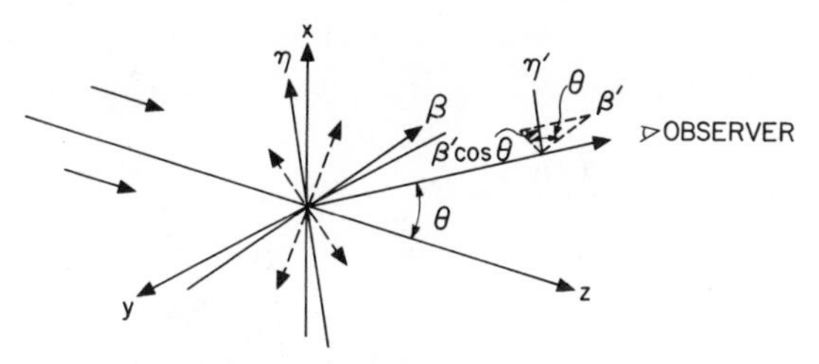

Fig. 10-2. A Rayleigh scattering diagram for explaining polarization of skylight.

η' component remains, and the scattered wave becomes completely polarized in the η direction. This simple explanation for the 90° complete polarization of skylight further demonstrated to Lord Rayleigh and others that skylight was clearly a scattering process. The polarization phenomenon is an interesting process to observe with a simple polarizer on a clear, bright day.

The Rainbow

One of the most interesting of scattering processes, the rainbow, cannot be explained by simple Rayleigh scattering theory. This striking visual phenomenon involves the interaction of rays with a spherical raindrop and the fact that the index of refraction of water is not constant with wavelength. The color spread can be seen from Snell's law,

$$\eta_1 \sin \theta_1 = \eta_2 \sin \theta_2 \tag{10-4}$$

where η_1 and θ_1 are associated with the external medium and η_2, which depends on λ, and η_2 are associated with the raindrop. From Eq. (10-4) it is apparent that θ_2 must depend on wavelength, thus giving an apparent spread in the spectrum of emerging rays shown in Fig. 10-3.

The second part of the rainbow explanation depends on the fact that, as the angle θ is varied, shown in Fig. 10-4(a) as rays 1, 2, or 3, it goes through a minimum, θ_0. This minimum condition is referred to as a stationary ray condition for one internal reflection and is shown as ray 2 in Fig. 10-4(a).

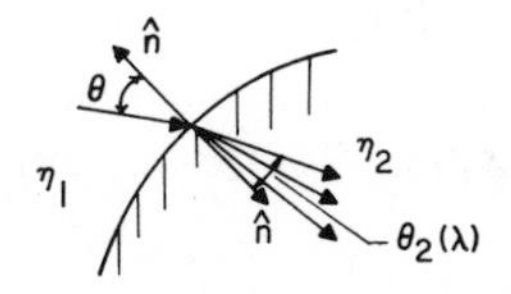

Fig. 10-3. Snell's law and curved surfaces in a dispersive media.

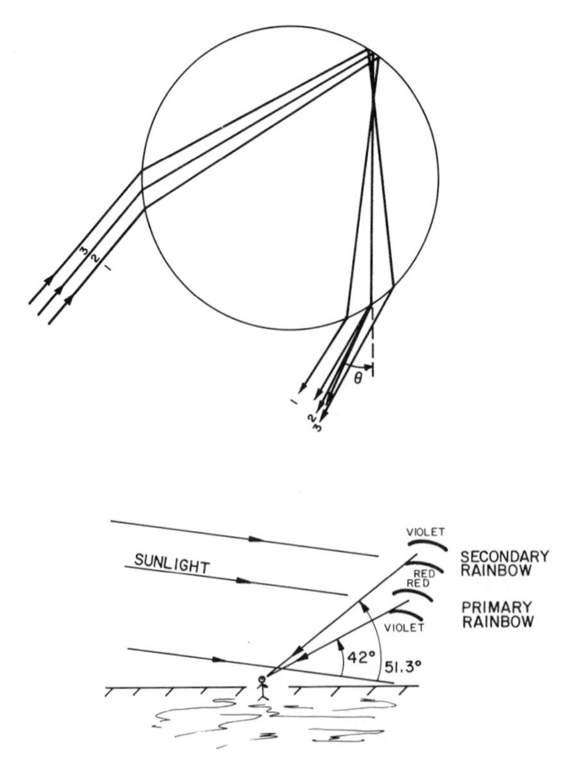

Fig. 10-4. (a) Formation of the rainbow in a spherical droplet. (b) Formation and relative position of primary and secondary rainbows.

Other minima can occur for more than one internal reflection. Two internal reflections produce a secondary rainbow. Figure 10-4(b) shows the details of how the rainbows are located with respect to the antisolar point. The stationary ray also has associated with it a brightening of intensity and dispersion. The variation of color about the minimum angle is about $\pm 1.5°$ with the bright violet appearing on the bottom. The secondary rainbow has one additional internal reflection, and thus has its colors inverted.

This first-order explanation of the rainbow is dependent on wave fronts being adequately characterized by their normals and local radii of curvature. This approximation breaks down for the minima associated with a rainbow. In this case, the next higher-order cubic wave-front approximation has to be used. This more detailed theory was developed by Airy [125], in which an amplitude scattering function $f(z)$ for any angle θ was derived

$$f(z) = \int_0^\infty \cos\{(\pi/2)(zt - t^3)\}\, dt \tag{10-5}$$

where t is the tangential distance along the wave front, normalized to a distance l that one can move in the tangential direction before encountering a phase shift of $\lambda/4$. The variable z is proportional to the change in angle about the minimum angle θ_0, multiplied by the ratio l divided by $\lambda/4$.

An even more detailed description of the rainbow and the associated scattering functions follows from the Mie theory for large particles ($ka > 2000$, a is the radius of the raindrop). In this case, the Mie scattering functions reduce to the Airy functions when tractable asymptotic forms are used. More will be presented on this after the section on Mie theory of scattering by spherical particles of arbitrary size.

Scattering by a Dielectric Sphere

The simple heuristic arguments given for scattering by a small dielectric sphere can be strengthened by a more rigorous development. The complete details are given in several texts [2, 4]. In this description, more of a summary will be given, highlighting the functional dependence of the scattering function $f_1(\eta_1, \eta_2)$.

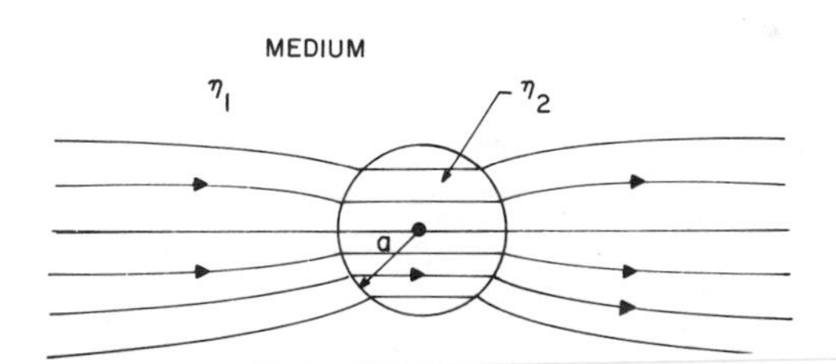

Fig. 10-5. Scattering and field lines of a dielectric sphere illuminated by a uniform electric field.

It is assumed that a linearly polarized field is incident on this small spherical particle ($a \ll \lambda$) such that across the particle a uniform electric field can be assumed to exist. This field polarizes the sphere to form dipoles with a resulting field distribution shown in Fig. 10-5. These dipoles arise for two reasons. One is the displacement of electronic charge, giving electronic polarizability, and a second term arising from alignment of existing permanent dipoles, giving orientational polarizability. The latter term is generally much smaller than the first term and is most often neglected.

The internal field can be shown† to be

$$\mathbf{E}_{\text{int}} = \left(\frac{3\eta_1^{\,2}}{\eta_2^{\,2} + 2\eta_1^{\,2}}\right)\mathbf{E}_0 \tag{10-6}$$

which results in an external dipole moment

$$\mathbf{P} = 4\pi\eta_1^{\,2}a^3\left(\frac{\eta_2^{\,2} - \eta_1^{\,2}}{\eta_2^{\,2} + 2\eta_1^{\,2}}\right)\mathbf{E}_0 \tag{10-7}$$

These dipoles oscillate synchronously with the exciting field and radiate secondary scattered fields. Further, the intensity‡ at a distance r from a scattering sphere is

$$I = \frac{16\pi^4 a^6}{r^2\lambda^4}\left(\frac{\eta_2^{\,2} - \eta_1^{\,2}}{\eta_2^{\,2} + 2\eta_1^{\,2}}\right)^2 \mid E_0 \mid^2 \sin\psi \tag{10-8}$$

where ψ is the angle between the dipole direction and the position vector $\mathbf{r}$ that locates the observer. Rewriting this intensity equation using the volume of the sphere ($V = \frac{4}{3}\pi a^3$) as a parameter gives

$$I = \left(\frac{I_0 V^2}{r^2\lambda^4}\right)a\pi^2\left(\frac{\eta_2^{\,2} - \eta_1^{\,2}}{\eta_2^{\,2} + 2\eta_1^{\,2}}\right)^2 \sin\psi \tag{10-9}$$

This result is consistent with the result given in Eq. (10-2), where the dependencies on V^2, r^{-2}, and λ^{-4} are all preserved, and substantiate the earlier heuristic derivation. This derivation gives considerable strength to the functional wavelength dependence and leads to an explicit representation $f_1(\eta_1, \eta_2)$. Equation (10-9) can be used directly to give a representation for the scattering function

$$f_1(\eta_1, \eta_2) = 9\pi^2\left(\frac{\eta_2^{\,2} - \eta_1^{\,2}}{\eta_2^{\,2} + 2\eta_1^{\,2}}\right)^2 \sin\psi \tag{10-10}$$

This result is applicable to sphere sizes that are constrained by $ka \leq 0.3$. It also is consistent with the original assertion that the scattering was due to perturbations in the refractive index. Thus, as η_2 approaches η_1, the scattering function decreases to zero. Another property is the angular dependence through the term $\sin\psi$. From this term the well-known null that lies along the direction of the dipole is seen. This also shows the maximum that lies along a direction perpendicular to the dipole or perpendicular to the exciting field direction. This then gives a strongly forward scattered wave.

† See J. A. Stratton [2], p. 436.

‡ See Stratton [2], p. 205.

A parameter which is quite useful is the scattering cross section. This term, when normalized by the geometric cross section of the sphere, leads quite naturally on the scattering factor q_{scat}. The scattering cross section can be obtained in terms of total scattered energy by integrating the scattered intensity over a total solid angle of 4π steradians.

The total scattered energy will be designated by

$$\Sigma_{\text{scat}} = \int_0^{\pi} \int_0^{2\pi} Ir^2 \sin \psi \, d\psi \, d\phi \tag{10-11}$$

where ϕ is the circumferential angle in the plane of the dipole. Performing the integration indicated in Eq. (10-11), the following result is obtained

$$\Sigma_{\text{scat}} = \frac{128\pi^5 a^6}{3\lambda^4} \left(\frac{\eta_2{}^2 - \eta_1{}^2}{\eta_2{}^2 + 2\eta_1{}^2} \right)^2 \tag{10-12}$$

The scattering factor q_{scat} is obtained by dividing by the geometric cross section of the sphere πa^2, leaving,

$$q_{\text{scat}} = \frac{8}{3} (ka)^4 \left(\frac{\eta_2{}^2 - \eta_1{}^2}{\eta_2{}^2 + 2\eta_1{}^2} \right)^2 \tag{10-13}$$

where $k = 2\pi/\lambda$ has been used for the wavelength dependence and also to give the result in terms of the classical size to wavelength ratio. The scattering factor represents the fraction of incident energy scattered in all directions.

This result will be useful when comparing with the Mie theory scattering result for arbitrary sized spheres. The original inverse fourth-power wavelength dependence is still clearly retained as plotted in Fig. 10-1.

This result is also quite general and applicable to the case of where absorption effects are also present. In the next section, absorption effects will be considered, and this scattering efficiency will be added to an equivalent absorption efficiency to give a total extinction effect.

Absorption Effects

The previous sections have all dealt with lossless scattering in dilute media. If absorption effects are considered, then the extinction is composed of two terms: the sum of scattering and absorption coefficients. In this case then, the Bouguer–Lambert extinction law [131] applies

$$I = I_0 e^{-\sigma' l} \tag{10-14}$$

where σ' represents a scattering coefficient per unit length and may have contributions from both the scattering and absorption cross sections.

Using the concept of transmission function developed in Chapter VI, we can define the transmission function τ as

$$\tau = I/I_0 = e^{-\sigma' x} \tag{10-15}$$

This function, τ, can be related to an equivalent optical density by writing

$$D(x) = \log_{10}(1/\tau) \tag{10-16}$$

which, when Eq. (10-15) is substituted, becomes

$$D(x) = \log_{10} e^{\sigma' x} = 0.4343\sigma' x \tag{10-17}$$

where $D(x)$ is measured in optical density units (o.d.). This particular form of Eq. (10-17) can be used to measure concentrations in absorbing layers. Knowledge of the thickness x, the specific absorption coefficients, and the respective concentrations is required for determining the transmission loss.

The relationship to Rayleigh scattering coefficients can be derived by noting that the decrease in intensity due to scattering can be expressed in differential form as

$$dI(x)/dx = -N\,\Sigma_{\text{scat}}\,I(x) \tag{10-18}$$

This equation can be solved by transposing $I(x)$ and dx and integrating. The boundary condition is that $I(0)$ equals I_0. This defines a scattering transmission function τ_s as

$$\tau_s = I_s(x)/I_0 = \exp[-tx] \tag{10-19}$$

where t is referred to as the turbidity of the medium [131] and is equal to $N\,\Sigma_{\text{scat}}$.

Since the effects of both transmission losses in a medium is multiplicative, the coefficients can be added giving a total extinction transmission function of

$$\tau_{\text{ext}} = \exp[-(t + \sigma')x] \tag{10-20}$$

where σ' represents an absorption coefficient. This result is useful in studies of biological cells and other dielectric media that have some absorption losses present.

Equation (10-20) is sometimes written in terms of an amplitude coefficient function γ such that direct correlations can be made with the imaginary part of the complex wave number. In this case, Eq. (10-20) becomes

$$\tau_{\text{ext}} = \exp[-(t + 2\gamma)x] \tag{10-21}$$

Mie–Debye Scattering Theory—Spherical Particles

The problem of electromagnetic wave scattering by spherical particles was an important one at the turn of the century. The terminology, Mie–Debye scattering theory, is used because both Mie [132] (1908) and Debye [133] (1909) solved essentially the same scattering problem. Rayleigh (1872) had done it for acoustic waves, but not for electromagnetic waves. The problem of scattering from a spherical dielectric particle of arbitrary size was one that could be formulated exactly and solved, at least in a formal sense, by considering a full-wave boundary-value problem. The formulation is simplified by using a spherical geometry and a special set of potential functions. The motivation for potential functions follows from the simplicity of the resulting equations and the ease of separation to the electric and magnetic effects (TE and TM modes).

The particular potential functions most useful in this problem are the electric Hertzian potential ψ and the magnetic Hertzian potential ψ^*. To obtain the **E** and **H** fields in terms of these simple scalar potential functions, one starts with the vector form of the Hertzian potentials $\boldsymbol{\pi}$ and $\boldsymbol{\pi}^*$ defined by[†]

$$\mathbf{E} = \nabla\nabla \cdot \boldsymbol{\pi} - \mu\varepsilon \frac{\partial^2 \boldsymbol{\pi}}{\partial t^2} - \mu\nabla \times \frac{\partial \boldsymbol{\pi}^*}{\partial t} \tag{10-22a}$$

and

$$\mathbf{H} = \varepsilon\nabla \times \frac{\partial \boldsymbol{\pi}}{\partial t} + \nabla\nabla \cdot \boldsymbol{\pi}^* - \mu\varepsilon \frac{\partial^2 \boldsymbol{\pi}^*}{\partial t^2} \tag{10-22b}$$

where $\boldsymbol{\pi}$ and $\boldsymbol{\pi}^*$ satisfy

$$\nabla^2 \boldsymbol{\pi} - \mu\varepsilon \frac{\partial^2 \boldsymbol{\pi}}{\partial t^2} = -\frac{\mathbf{P}_0}{\varepsilon_0} \tag{10-23a}$$

and

$$\nabla^2 \boldsymbol{\pi}^* - \mu\varepsilon \frac{\partial^2 \pi^*}{\partial t^2} = -\mathbf{M}_0 \tag{10-23b}$$

The polarizations $\mathbf{P}_0$ and $\mathbf{M}_0$ represent the fixed electric and magnetic dipoles of the medium. Thus from Eqs. (10-23) it is clear that the electric and magnetic effects are separated because of the separated source terms. To complete the definition of the Hertzian vector potentials, the solenoidal part of each vector field must be defined. Since the region is assumed to be source free, the divergence of $\boldsymbol{\pi}$ and $\boldsymbol{\pi}^*$ can be defined by ψ and ψ^* as

$$\nabla \cdot \boldsymbol{\pi} = -\psi \tag{10-24a}$$

† See Stratton [2], p. 29.

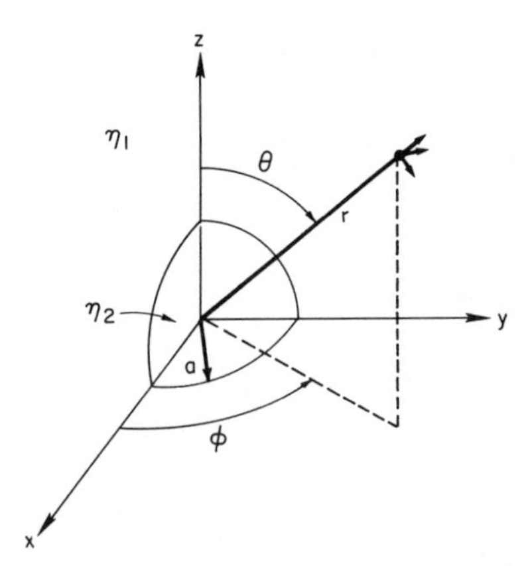

Fig. 10-6. Spherical coordinate system for scattering by a spherical dielectric particle of a radius a and index of refraction η_2.

and

$$\nabla \cdot \boldsymbol{\pi}^* = -\psi^* \tag{10-24b}$$

where ψ and ψ^* both satisfy the scalar wave equation. Substitution of Eqs. (10-24) and expansion of Eqs. (10-22) provide the vector components of **E** and **H** in terms of the scalar Hertzian potentials† in the spherical coordinate system shown in Fig. 10-6.

$$E_r = \frac{\partial^2(r\psi)}{\partial r^2} + k^2 r\psi \tag{10-25a}$$

$$E_\theta = \frac{1}{r}\,\frac{\partial^2(r\psi)}{\partial r\,\partial\theta} + \frac{j\omega\mu}{r\sin\theta}\,\frac{\partial(r\psi^*)}{2\phi} \tag{10-25b}$$

$$E_\phi = \frac{1}{r\sin\theta}\,\frac{\partial^2(r\psi)}{\partial r\,\partial\phi} - \frac{j\omega\mu}{r}\,\frac{\partial(r\psi^*)}{\partial\theta} \tag{10-25c}$$

$$H_r = \frac{\partial^2(r\psi^*)}{\partial r^2} + k^2 r\psi^* \tag{10-25d}$$

$$H_\phi = \frac{-j\omega\varepsilon}{r\sin\theta}\,\frac{\partial(r\psi)}{\partial\phi} + \frac{1}{r}\,\frac{\partial^2(r\psi^*)}{\partial r\,\partial\theta} \tag{10-25e}$$

$$H_\theta = \frac{j\omega\varepsilon}{r}\,\frac{\partial(r\psi)}{\partial\theta} + \frac{1}{r\sin\theta}\,\frac{\partial^2(r\psi^*)}{\partial r\,\partial\phi} \tag{10-25f}$$

where k^2 is wave number defined in Eq. (1-11).

† See Born and Wolf [11], p. 638.

Now since ψ and ψ^* satisfy scalar-type wave equations, Eq. (1-13) needs to be expanded in the spherical coordinate system, and then solved by the separation of variables technique. For ψ, the expanded wave equation becomes

$$\frac{1}{r}\left[\frac{\partial^2(r\psi)}{\partial r^2}\right] + \frac{1}{r^2 \sin\theta}\frac{\partial}{\partial\theta}\left(\sin\theta\frac{\partial\psi}{\partial\theta}\right) + \frac{1}{r^2\sin^2\theta}\frac{\partial^2\psi}{\partial\phi^2} + k^2\psi = 0 \tag{10-26}$$

A similar equation can be obtained for ψ^* by direct substitution of ψ^* for ψ. The method of separation of variables assumes that $\psi(r, \theta, \phi)$ can be represented as a product of three functions, each dependent on a single coordinate direction,

$$\psi(r, \theta, \phi) = R(r)\Theta(\theta)\Phi(\phi) \tag{10-27}$$

Substitution of Eq. (10-27) into Eq. (10-26) results in three well-known ordinary differential equations†

$$\frac{d^2[rR(r)]}{dr^2} + \left[k^2 - \frac{n(n+1)}{r^2}\right]rR(r) = 0 \tag{10-28a}$$

$$\frac{1}{\sin\theta}\frac{d}{d\theta}\left[\sin\theta\frac{d\Theta(\theta)}{d\theta}\right] + \left[n(n+1) - \frac{m^2}{\sin^2\theta}\right]\Theta(\theta) = 0 \tag{10-28b}$$

and

$$\frac{d^2\Phi(\phi)}{d\phi^2} + m^2\Phi(\phi) = 0 \tag{10-28c}$$

where n is an integer and m can assume the values such that $|m| < n$.

The solution of the azimuthal equation, Eq. (10-28c), is a well-known result

$$\Phi(\phi) = a_m \cos m\phi + b_m \sin m\phi \tag{10-29}$$

where a constraint of single valuedness has been imposed from the constraints on the actual physical fields.

The solution for the polar angle equation results in the special set of functions known as the associated Legendre polynomials or associated Legendre functions

$$\Theta(\theta) = P_n^{(m)}(\cos\theta) \tag{10-30}$$

These functions vanish identically if $|m| > n$. Because of the interrelatedness of m and n, there are $2n + 1$ such functions for each n.

† See Born and Wolf [11], p. 639.

The solutions for the radial equation, Eq. (10-28a), can be obtained by rewriting the equation in a standard Bessel form. Let $Z(kr)$ be a generalized Bessel function defined by

$$(1/\sqrt{kr})Z(kr) = R(r) \tag{10-31}$$

which, when substituted into Eq. (10-28a), results in a new radial equation

$$\frac{d^2Z(kr)}{d(kr)^2} + \frac{1}{kr}\frac{dZ(kr)}{d(kr)} + \left[1 - \frac{(n+1/2)^2}{(kr)^2}\right]Z(kr) = 0 \tag{10-32}$$

The solution to this equation is a standard cylindrical function $Z_{n+1/2}(kr)$ of order $(n + 1/2)$. Therefore, the solution for $R(r)$ is

$$R(r) = (1/\sqrt{kr})Z_{n+1/2}(kr) \tag{10-33}$$

Since Eq. (10-28a) is a second-order differential equation, then $R(r)$ is expressed as a linear combination of the two possible Bessel functions representable by $Z_{n+1/2}(kr)$. These two functions are the standard Bessel function $J_{n+1/2}(kr)$ and the Neumann function $N_{n+1/2}(kr)$. Thus $R(r)$ is representable as linear combination of two functions $\xi_n(kr)$ and $\chi_n(kr)$ as

$$rR(r) = c_n\xi_n(kr) + d_n\chi_n(kr) \tag{10-34}$$

where

$$\xi_n(kr) = (\pi kr/2)^{1/2}J_{n+1/2}(kr) \tag{10-35}$$

and

$$\chi_n(kr) = -\sqrt{\pi kr/2}\,N_{n+1/2}(kr) \tag{10-36}$$

The boundedness of $J(\ )$ and $N(\ )$ will be important inside the sphere where $N(\ )$ diverges as $r \to 0$ and $J_n(\ )$ is finite. The constants c_n and d_n for outside the sphere can be determined from the radiation condition; namely, $Z_n(kr)$ functions or linear combinations thereof must converge to zero as $r \to \infty$. This constraint† is satisfied by a Hankel function of the first kind [2], $H^{(1)}_{n+1/2}(kr)$. Therefore the radial function outside the sphere is‡

$$rR_o(r) = \zeta_n^{(1)}(kr) = \xi_n(kr) - j\chi_n(kr) = (\pi kr/2)^{1/2}H^{(1)}_{n+1/2}(kr) \tag{10-37}$$

† Assuming a time dependence of $e^{-j\omega t}$.

‡ The Bessel, Neumann, and Hankel functions are related by relationships that are like the Euler relations for exponential, sine, and cosine functions, namely,

$$J_n + jN_n = H_n^{(1)}; \qquad J_n - jN_n = H_n^{(2)}$$

The general solution of the scalar wave equation in spherical coordinates can now be expressed as a double sum of products, as

$$r\psi = r \sum_{n=0}^{\infty} \sum_{m=-n}^{n} \{c_n \xi_n(kr) + d_n \chi(kr)\}\{P_n^{(m)}(\cos\theta)\} \times \{a_m \cos m\phi + b_m \sin m\phi\} \tag{10-38}$$

This general potential function, written for both electric and magnetic cases, is used in the general equations, Eqs. (10-26), to find the various **E** and **H** vector components. This is done in order that the boundary condition of continuity of fields can be applied.

However, before the continuity conditions can be applied, the incident field has to be expressed in spherical harmonics. It is assumed that an electric field, of unit amplitude, plane polarized along the x direction is propagating in the positive z direction, like

$$E_x = e^{jk_1 z} \tag{10-39}$$

where $k_1 = \eta_1 k_0$. Since the incident wave must be finite at the origin, expansion of the two incident Hertz potentials results in[†]

$$r\psi_{\mathrm{i}} = \frac{1}{k_1^2} \sum_{n=1}^{\infty} j^{n-1} \frac{(2n+1)}{n(n+1)} \xi_n(k, r) P_n^{(1)}(\cos\theta) \cos\phi \tag{10-40}$$

and

$$r\psi_{\mathrm{i}}^* = \frac{1}{k_1^2} \sum_{n=1}^{\infty} j^{n-1} \frac{(2n+1)}{n(n+1)} \xi_n(k, r) P_n^{(1)}(\cos\theta) \sin\phi \tag{10-41}$$

where $P_n^{(1)}(\cos\theta)$ are the associated Legendre functions of the first kind [2, 11].

The boundary conditions of continuity on the tangential fields can be translated to boundary conditions on the ψ and ψ^* functions [4, 11]

$$(\partial/\partial r)\{r(\psi_{\mathrm{i}} + \psi_{\mathrm{s}})\}_{r=a} = (\partial/\partial r)\{r\psi_d\}_{r=a} \tag{10-42a}$$

$$(\partial/\partial r)\{r(\psi_{\mathrm{i}}^* + \psi_{\mathrm{s}}^*)\}_{r=a} = (\partial/\partial r)\{r\psi_d^*\}_{r=a} \tag{10-42b}$$

$$\eta_2^2(\psi_{\mathrm{i}} + \psi_{\mathrm{s}})_{r=a} = \eta_1^2 \psi_d \,|_{r=a} \tag{10-42c}$$

$$(\psi_{\mathrm{i}}^* + \psi_{\mathrm{s}}^*)_{r=a} = \psi_d^* \,|_{r=a} \tag{10-42d}$$

where the medium has been assumed to have constant permeability. Since

† See Born and Wolf [11], p. 641.

these equations are linear independent equations, they must apply to each corresponding term in the expansions. Thus the following four sets of equations result

$$(\eta_2/\eta_1)[\xi_n'(k_1a) - a_n\zeta_n'(k_1a)] = c_n\xi_n'(k_2a) \tag{10-43a}$$

$$(\eta_2/\eta_1)[\xi_n'(k_1a) - b_n\zeta_n'(k_1a)] = d_n\xi_n'(k_2a) \tag{10-43b}$$

$$\xi_n(k_1a) - a_n\zeta_n(k_1a) = c_n\xi_n(k_2a) \tag{10-43c}$$

$$(\eta_2/\eta_1)^2[\xi_n(k_1a) - b_n\zeta_n(k_1a)] = d_n\xi_n(k_2a) \tag{10-43d}$$

where the prime on the functions indicates differentiation with respect to the argument of the function. Equations (10-43) can be solved for the scattered field coefficients a_n and b_n, yielding

$$a_n = \frac{\xi_n(k_1a)\xi_n'(k_2a) - (\eta_2/\eta_1)\xi_n(k_2a)\xi_n'(k_1a)}{\zeta_n(k_1a)\xi_n'(k_2a) - (\eta_2/\eta_1)\xi_n(k_2a)\zeta_n'(k_1a)} \tag{10-44a}$$

$$b_n = \frac{(\eta_2/\eta_1)\xi_n(k_1a)\xi_n'(k_2a) - \xi_n(k_2a)\xi_n'(k_1a)}{(\eta_2/\eta_1)\zeta_n(k_1a)\xi_n'(k_2a) - \xi_n(k_2a)\zeta_n'(k_1a)} \tag{10-44b}$$

Because the generalized Bessel functions are convergent to zero for large argument ($kr \gg n$), at least when compared to the order, the functions can be replaced with asymptotic forms. In particular, the Hankel functions can be represented in a simple exponential form

$$\zeta_n(k_1r) = j^{n+1}\exp[jk_1r] \tag{10-45a}$$

and

$$\zeta_n'(k_1r) = j^n \exp[jk_1r] \tag{10-45b}$$

For the far field, the scattered waves have only transverse components. This can be seen by noting that the radial fields decay like r^{-2}, where the transverse components decay like r^{-1}. Further, the transverse H_θ and H_ϕ can be related to the orthogonal transverse E_ϕ and E_θ by simple impedance relationships. Thus using the approximations outlined above, the transverse fields are

$$E_\phi = \frac{H_\theta}{\eta_1} = \frac{j\exp[jk_1r]}{k_1r}\sin\phi\sum_{n=1}^{\infty}\frac{2n+1}{n(n+1)} \times \left\{a_n\frac{P_n^{(1)}(\cos\theta)}{\sin\theta} - b_n\frac{d(P_n^{(1)}(\cos\theta)}{d\theta}\right\} \tag{10-46a}$$

and

$$E_\theta = -\frac{H_\phi}{\eta_1} = -\frac{j\exp[jk_1 r]}{k_1 r}\cos\phi \sum_{n=1}^{\infty} \frac{2n+1}{n(n+1)} \times \left\{ a_n \frac{dP_n^{(1)}(\cos\theta)}{d\theta} - b_n \frac{P_n^{(1)}(\cos\theta)}{\sin\theta} \right\} \tag{10-46b}$$

The inverse dependence on r shows that these waves are like spherical waves, which is consistent with the notion that they arise from sources within the scattering region.

It is convenient to simplify the results at this point by defining amplitude functions. This is important when it is desired to describe the in-plane intensity I_θ and the out-plane intensity I_ϕ, which is perpendicular to the scattering plane. These two intensities are from similarly polarized incident components, except that there is an angular modulation function $\sin^2\phi$ for $I_\phi^{(0)}$ incident, and $\cos^2\phi$ for $I_\theta^{(0)}$ incident. The amplitude functions A_1 and A_2 are defined by

$$A_1 = \sum_{n=1}^{\infty} \frac{2n+1}{n(n+1)} \left\{ a_n \frac{P_n^{(1)}(\cos\theta)}{\sin\theta} + b_n \frac{d[P_n^{(1)}(\cos\theta)]}{d\theta} \right\} \tag{10-47a}$$

and

$$A_2 = \sum_{n=1}^{\infty} \frac{2n+1}{n(n+1)} \left\{ a_n \frac{d[P_n^{(1)}(\cos\theta)]}{d\theta} + b_n \frac{P_n^{(1)}(\cos\theta)}{\sin\theta} \right\} \tag{10-47b}$$

The in-plane and out-plane intensities are given in simple form in terms of A_1 and A_2 as

$$I_\theta = \frac{|A_2|^2}{k_1{}^2 r^2}\sin^2\phi = \frac{\lambda^2}{4\pi^2 r^2}|A_2|^2 \sin^2\phi \tag{10-48a}$$

and

$$I_\phi = \frac{|A_1|^2}{k_1{}^2 r^2}\cos^2\phi = \frac{\lambda^2}{4\pi^2 r^2}|A_1|^2 \cos^2\phi \tag{10-48b}$$

In general, there is a phase difference δ between these two components given by

$$\tan\delta = \frac{\mathrm{Re}\{A_1\}\,\mathrm{Im}\{A_2\} - \mathrm{Re}\{A_2\}\,\mathrm{Im}\{A_1\}}{\mathrm{Re}\{A_1\}\,\mathrm{Re}\{A_2\} + \mathrm{Im}\{A_1\}\,\mathrm{Im}\{A_2\}} \tag{10-49}$$

This means that in general the scattered light is elliptically polarized. Also if the incident radiation is unpolarized with unit magnitude, then the scattered radiation intensity is

$$I_u = \frac{\lambda^2}{8\pi^2 r^2}(|A_1|^2 + |A_2|^2) \tag{10-50}$$

where the degree of polarization of this radiation is given by

$$P = \left| \frac{|A_1|^2 - |A_2|^2}{|A_1|^2 + |A_2|^2} \right| \tag{10-51}$$

Mie–Debye Cross Sections and Efficiency Factors

The scattering cross sections can be evaluated by noting that the extinction cross section will be composed of two parts: the scattering and absorption components, represented by

$$\Sigma_{\text{ext}} = \Sigma_{\text{scat}} + \Sigma_{\text{abs}} \tag{10-52}$$

Extinction then corresponds to the total energy abstracted from the incident beam. To evaluate the total outward flow of energy, the real part, Re{ }, of the Poynting vector integrated over the total solid angle of the bounding surface is used. The bounding sphere as a radius r, large compared to the scattering region volume size ($k_1 r \gg k_1 \sqrt[3]{V}$). The time-averaged radial Poynting vector is

$$\begin{aligned} S_r &= S_{1r} + S_{2r} + S_{3r} \\ &= \tfrac{1}{2}\{E_{i\theta}H_{i\phi}^* - E_{i\phi}H_{i\theta}^*\} + \tfrac{1}{2}\{E_{s\theta}H_{i\phi}^* + E_{i\theta}H_{s\phi}^* - E_{s\phi}H_{i\theta}^* - E_{i\phi}H_{s\theta}^*\} \\ &\quad + \tfrac{1}{2}\{E_{s\theta}H_{s\phi}^* - E_{s\phi}H_{s\theta}^*\} \end{aligned} \tag{10-53}$$

The real part of this time-averaged Poynting vector integrated over the bounding sphere must be physically equal to the minus of the absorbed energy in the scattering sphere; therefore

$$-\Sigma_{\text{abs}} = \text{Re}\left\{\int_0^\pi \int_0^{2\pi} S_r r^2 \sin\theta \, d\theta \, d\phi\right\} \tag{10-54a}$$

$$\begin{aligned} &= \text{Re}\left\{\int_0^\pi \int_0^{2\pi} S_{1r} r^2 \sin\theta \, d\theta \, d\phi\right\} + \text{Re}\left\{\int_0^\pi \int_0^{2\pi} S_{2r} r^2 \sin\theta \, d\theta \, d\phi\right\} \\ &\quad + \text{Re}\left\{\int_0^\pi \int_0^{2\pi} S_{3r} r^2 \sin\theta \, d\theta \, d\phi\right\} \end{aligned} \tag{10-54b}$$

The first term, corresponding to S_{ir}, has to be zero since the conductivity of the sphere is zero and the net incident energy flux is zero over the bounding sphere. The last term is the scattered energy which is Σ_{scat}. Therefore, transposing the terms, we note that the second term is

$$\text{Re}\left\{\int_0^\pi \int_0^{2\pi} S_{2r} r^2 \sin\theta \, d\theta \, d\phi\right\} = \Sigma_{\text{scat}} + \Sigma_{\text{abs}} \tag{10-55}$$

but using Eq. (10-52) this is Σ_{ext}; therefore

$$\Sigma_{\text{ext}} = \text{Re}\left\{\int_0^{\pi} \int_0^{2\pi} S_{2r} r^2 \sin\theta \, d\theta \, d\phi\right\} \tag{10-56}$$

These integrals were evaluated by Mie [132] (1908), Debye [133] (1909), and others leading to

$$\Sigma_{\text{scat}} = \left(\frac{\lambda^2}{2\pi}\right) \sum_{n=1}^{\infty} (2n+1)\{|a_n|^2 + |b_n|^2\} \tag{10-57}$$

and

$$\Sigma_{\text{ext}} = \left(\frac{\lambda^2}{2\pi}\right) \sum_{n=1}^{\infty} (2n+1)\{\text{Re}(a_n + b_n)\} \tag{10-58}$$

which can be compared with the Rayleigh results if a_n and b_n are substituted from Eqs. (10-44).

The corresponding efficiency factors can be derived by dividing Eqs. (10-58) and (10-57) by the geometric cross section of the sphere πa^2, leaving

$$q_{\text{scat}} = \frac{2}{(k_1 a)^2} \sum_{n=1}^{\infty} (2n+1)\{|a_n|^2 + |b_n|^2\} \tag{10-59}$$

and

$$q_{\text{ext}} = \frac{2}{(k_1 a)^2} \sum_{n=1}^{\infty} (2n+1)\{\text{Re}(a_n + b_n)\} \tag{10-60}$$

Notice that these results, Eqs. (10-57) through (10-60), are independent of the polarizations.

These cross section and efficiency factor results are important when comparing various media and when attempting to evaluate expected attenuations in a given medium. Many authors [125, 131] have used these results in describing the effects of rain and other atmospheric attenuation effects. Limiting or asymptotic forms of these field expressions and cross sections have been used to explain such effects as rainbows† and the glory [120] phenomena.

Considerable work [4, 11] has been done to explain the terms of the field expressions as multipole [2] source terms. In the far field one can think of equivalent sources that are superposed and made of very complex dipole, quadrapole, and other complex multipole structures [11]. Born and Wolf [11] and Stratton [2] describe these waves as partial waves and give pictorial representations how each wave mode might be distributed, much

† See Van de Hulst [125], Section 12.35, for further details and expansions. The application of these higher-order expansions was given in the section on the rainbow.

as one would envision the TM and TE field distributions in waveguides and cavities.

The above theory is primarily related to spherically shaped bodies. For other arbitrarily shaped dielectric bodies other methods are used [134]. These methods are primarily numerical and are most useful near the resonance region ($\lambda \simeq l$).

The Correspondence between Mie and Rayleigh Scattering

One of the major results of the Mie scattering theory was in the removal of the size restriction in the explanation of scattering phenomena, making it possible to understand a broad range of scattering effects. In addition, a correspondence was established with the existing Rayleigh theory by examination of the Mie–Debye theory at small ka. Small ka corresponds to the case of $a \ll \lambda$, but broadens the notion to a much larger wavelength region. Thus the scattering effects of particles in the microwave region [135] can be described by the inverse fourth-power relationship (Rayleigh scattering) if $ka < 0.3$.

If the scattering efficiency Q_{scat} is plotted versus ka, as shown in Fig. 10-7, the effect of the inverse wavelength-size ratio can be understood quite simply. Notice that for $ka < 0.3$, the inverse fourth-power dependence on wavelength is preserved. For $ka > 0.3$, the scattering is much less dependent on wavelength, and is in fact asymptotic to a constant for large ka.

This behavior is very important in explaining some phenomena such as white clouds. Here almost all wavelengths are scattered with roughly the

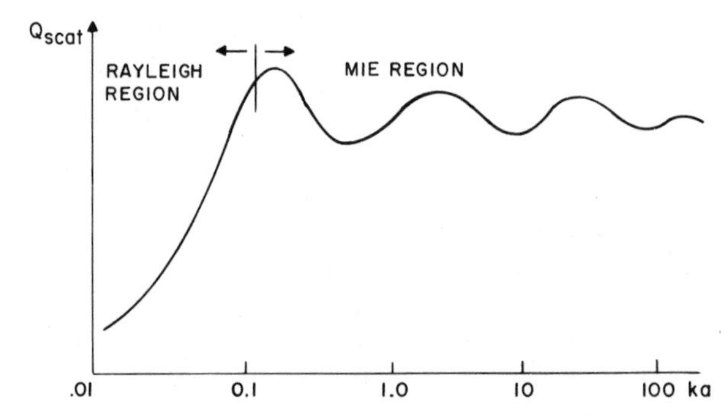

Fig. 10-7. Scattering efficiency Q_{scat} versus ka covering the Rayleigh and Mie regions of scattering.

same efficiency, if the drops are big enough.† Actual numbers for Q_{scat} versus ka have been generated [136] for refractive indices in the range of 1.33 to 1.50. The curves are very similar to Fig. 10-7 with the asymptotic Q_{scat} at large ka being about two.

Scattering in Random Media

The scattering processes discussed in the first part of this chapter are generally related to considering the scatterers as discrete or deterministic in shape and character. Very often it is not possible to model the scatterers in this fashion, and it is then necessary to treat the problem as scattering by random perturbations or changes in the medium [5, 138, 139]. Examples of such phenomena are the twinkling of stars, the shimmering of heat over

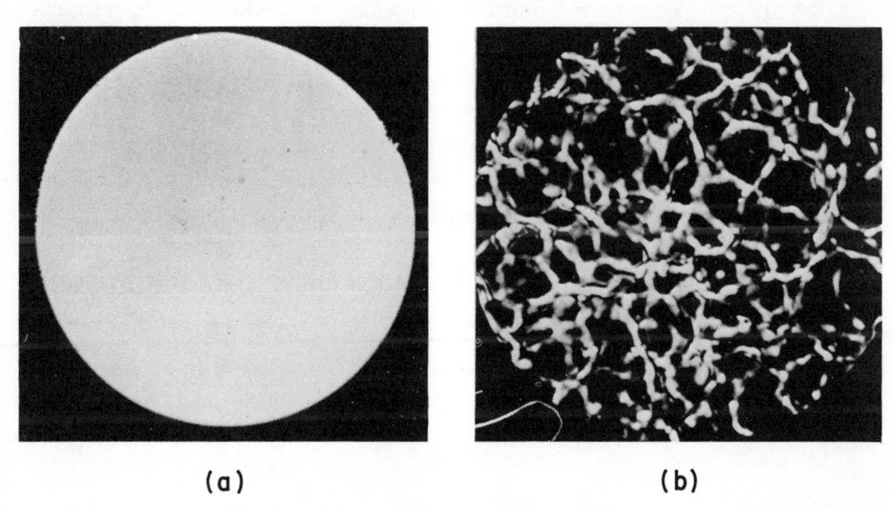

Fig. 10-8. Scattering of an optical beam in a random media. (a) Input beam at transmitting aperture. (b) Output beam at receiving aperture located 300 m down range.

a hot road, fluctuations in transhorizon radio communications, image jitter in a telescope, or more recently in optical communications paths. An example of what happens to an optical beam when it propagates just a few hundred meters in a near-earth atmosphere is shown in Fig. 10-8. Figure

† This argument is excluding the multiple scattering effects that generally occur in dense media like clouds. However, multiple scattering [135, 137] further reduces the wavelength dependence.

10-8(a) shows the beam at the beginning of the path. Figure 10-8(b) shows the severe corruption of the beam at the end of the path.

The fluctuations in the refractive index that cause this spatial modulation effect are due primarily to temperature fluctuations [5]. For optical waves this relationship is given by [126]

$$\Delta N = \Delta\eta \times 10^6 = -77.6 P_{\mathrm{DA}} (\Delta T / T_0^2) \tag{10-61}$$

where η is the refractive index, P_{DA} is the partial pressure of dry air in millibar, T is the temperature in degrees Kelvin, and N is defined by

$$N = (\eta - 1) \times 10^6 \tag{10-62}$$

These fluctuations originate from large-scale phenomena on the order of hundreds of meters and are then broken down and mixed by the wind until almost all scale sizes from hundreds of meters to several millimeters contribute to the fluctuation phenomena. The contribution of each size is not the same but follows a distribution, shown in Fig. 10-9, and which will be discussed later. The important point at this stage is that distances over which a variation exists are very much larger than a wavelength and, in general, do not vary with position in an average or mean sense. Thus, refractive effects will be neglected, and only the scattering from the random fluctuations will be considered.

For the random refractive index fluctuation model, the refractive index

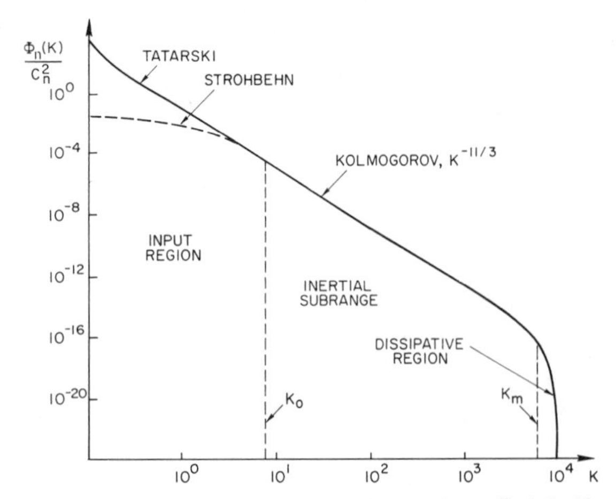

Fig. 10-9. Spectrum of the refractive index fluctuations $\Phi_n(K)$ [from Strohbehn (1971)].

of the atmosphere can be written as [5, 126]

$$\eta(\mathbf{r}, t) = \eta_0 + \eta_1(\mathbf{r}, t) \tag{10-63}$$

where η_0 is the average refractive index, $\langle\eta(\mathbf{r}, t)\rangle$, which in the atmosphere is generally very close to unity. Since most temperature fluctuations and wind speeds are very slow compared to a period of an optical wave, the time variation in Eq. (10-63) can be suppressed. In another sense, the actual time fluctuations can be thought of as separate realizations of the random field.

The scattering problem now reduces to deriving the appropriate wave equation for the case where the refractive index depends on position. Referring to Eq. (1-7)

$$\nabla\times\nabla\times\mathbf{E} = \nabla(\nabla\cdot\mathbf{E}) - \nabla^2\mathbf{E} = k^2\eta\mathbf{E} \tag{10-64}$$

it is seen that the difference in this derivation is the grad-divergence term, $\nabla(\nabla\cdot\mathbf{E})$. For a charge-free region, the divergence equation that is zero is the divergence of the displacement vector $\nabla\cdot(\varepsilon\mathbf{E})$. Expanding the divergence of the $\varepsilon\mathbf{E}$ term gives

$$\nabla\cdot\mathbf{E} = \mathbf{E}\cdot\nabla(\ln\varepsilon) \tag{10-65}$$

Thus the wave equation that results from Eq. (10-64) and Eq. (10-65) is

$$\nabla^2\mathbf{E} + k^2\eta E = -\nabla[\mathbf{E}\cdot\nabla(\ln\varepsilon)] \tag{10-66}$$

where the variation in ε becomes a source term in the basic representation of the field in the medium. If Eq. (10-63) is squared and substituted into Eq. (10-66) for ε, the following equation results

$$\nabla^2\mathbf{E} + k^2\mathbf{E} = -2\nabla(\mathbf{E}\cdot\nabla[\ln(1+\eta_1)] - 2k^2\eta_1\mathbf{E} \tag{10-67}$$

where variations on the order of $\eta_1{}^2$ have been neglected.

From the work with Green's functions in Chapter I, it appears that the solution to Eq. (10-67) could be written as an integral equation over the source term. This would, however, not be a solution in the strict sense since it is merely a translation from a differential equation to an integral equation.

Thus one is faced with a choice of method in solving Eq. (10-67). Generally, the simpler techniques should be explored first. In the case that the fluctuations or perturbations in η are weak, a perturbation technique on the fields can be used. The method of small perturbations is used to seek a solution in the form

$$\mathbf{E} = \mathbf{E}_0 + \mathbf{E}_1 + \mathbf{E}_2 + \cdots \tag{10-68}$$

where the $\mathbf{E}_j$ term has an order of smallness similar to $\eta_1{}^j$.† Substitution of Eq. (10-68) into Eq. (10-67) and then separation by equating like powers of smallness yield

$$\nabla^2\mathbf{E}_0 + k^2\mathbf{E}_0 = 0 \tag{10-69a}$$

$$\nabla^2\mathbf{E}_1 + k^2\mathbf{E}_1 = -2k^2\eta_1 E_0 - 2\nabla[\mathbf{E}_0 \cdot \nabla(\ln \eta_1)] \tag{10-69b}$$

where terms smaller than $\eta_1{}^2$ have been neglected. Physically, $\mathbf{E}_0$ represents the primary wave in the scattering process. The term $\mathbf{E}_1$ represents the scattered wave term and arises from the source terms that are directly proportional to $\mathbf{E}_0$.

Now it is appropriate to apply the integral equation representation [5, 66] and write the solution for one component of $\mathbf{E}_1$ as

$$E_1(\mathbf{r}') = \frac{k^2}{2\pi}\iiint_V \frac{\eta_1(\mathbf{r})E_0(\mathbf{r})\exp(jk\,|\,\mathbf{r}'-\mathbf{r}\,|)}{|\,\mathbf{r}'-\mathbf{r}\,|}\,d\mathbf{r} + \frac{1}{2\pi}\iiint_V \frac{\nabla[\mathbf{E}_0(\mathbf{r})\cdot\nabla\ln\eta_1(\mathbf{r})]\exp(jk\,|\,\mathbf{r}'-\mathbf{r}\,|)}{|\,\mathbf{r}'-\mathbf{r}\,|}\,d\mathbf{r} \tag{10-70}$$

where V is the scattering volume, $\mathbf{r}$ and $\mathbf{r}'$ represent an ordered triplet, and the origin of the coordinates is in the scattering volume.

If the scattering volume is characterized by a linear dimension that is very small compared to the distance to the observer, and the condition $\lambda r' \gg V^{2/3}$ is satisfied, then a Fraunhofer approximation can be implemented, leading to

$$E_1(\mathbf{r}') = \frac{k^2}{2\pi}\frac{e^{jkr'}}{r'}\iiint_V \eta_1(\mathbf{r})E_0(\mathbf{r})\exp\left(-j\frac{k\mathbf{r}\cdot\mathbf{r}'}{r'}\right)d\mathbf{r}' + \frac{1}{2\pi}\frac{e^{jkr'}}{r'}\iiint_V \nabla[\mathbf{E}_0(\mathbf{r})\cdot\nabla\ln\eta_1(\mathbf{r})]\exp\left(-j\frac{k\mathbf{r}\cdot\mathbf{r}'}{r'}\right)d\mathbf{r}' \tag{10-71}$$

where r' represents the magnitude of the distance from scattering volume to the observer.‡

† In the atmosphere η_1 may vary over the range of 10^{-4} to 10^{-6}. Thus not many terms are required.

‡ Note that in the Fraunhofer-type approximation it is implicit that $|\,\mathbf{r}'-\mathbf{r}\,| \sim r'$.

If one assumes a transverse wave like a plane wave for the primary wave $\mathbf{E}_0(\mathbf{r}')$, then the second term on the right-hand side of Eq. (10-71) can be shown to be canceled by the longitudinal component of the first term on the right-hand side.† Thus the scattered field is purely a transverse field in the Fraunhofer zone. The scattered field can be simply represented by the first term of Eq. (10-71). In this form the solution is known as the Born approximation.

As in other scattering theories the parameter of interest is the flow of scattered energy. Through the use of the Poynting theorem [1], the scattered energy density for the plane-wave case can be calculated. It is, however, simpler to examine the scattered intensity, which can be written as [5]

$$I(\mathbf{r}') = \frac{k^4 A_0{}^2 \sin^2 \psi}{4\pi^2 r'^2} \iiint_{V_1} d\mathbf{r}_1 \iiint_{V_2} d\mathbf{r}_2\, \eta_1(\mathbf{r}_1)\eta_1(\mathbf{r}_2) \times \exp\left[j\left(\mathbf{k} - \frac{k\mathbf{r}'}{r'}\right)\cdot(\mathbf{r}_1 - \mathbf{r}_2)\right] \tag{10-72}$$

where a plane wave, $\mathbf{A}_0 e^{j\mathbf{k}\cdot\mathbf{r}}$, has been assumed for $\mathbf{E}_0(\mathbf{r})$. The angle ψ is the angle between the polarization of incident wave field $\mathbf{E}_0$ and the vector from the scattering volume to the observer at $\mathbf{r}'$. It is interesting to note for this form of the scattered intensity that the fourth-power dependence on inverse λ is preserved in the k^4 term. In addition, the inverse length squared dependence is also explicit.

The intensity represented by Eq. (10-72) is a random quantity which has usefulness in terms of the measures or moments imposed on this random variable. One measure that is very important is the mean value, designated by an overbar

$$\overline{I(\mathbf{r}')} = \frac{k^4 A_0{}^2 \sin^2 \psi}{4\pi^2 r'^2} \iiint_{V_1} d\mathbf{r}_1 \iiint_{V_2} d\mathbf{r}_2 \overline{\eta_1(\mathbf{r}_1)\eta_1(\mathbf{r}_2)} \times \exp\left[j\left(\mathbf{k} - \frac{k\mathbf{r}'}{r'}\right)\cdot(\mathbf{r}_1 - \mathbf{r}_2)\right] \tag{10-73}$$

where a shorthand notation on the triple volume integrals has been used, and the overbar on the product $\overline{\eta_1(\mathbf{r}_1)\eta_1(\mathbf{r}_2)}$ represents a correlation function on the refractive index fluctuations.

† For this proof, use is made of [140]

$$\int_V \phi\nabla\psi\, d\mathbf{r} = \int_V \nabla(\phi\psi)\, d\mathbf{r} - \int_V \psi\nabla\phi\, d\mathbf{r} = \int_S \phi\psi\hat{n}\, ds - \int_V \psi\nabla\phi\, d\mathbf{r}$$

The correlation function on η_1 is a second-order measure on the medium fluctuations. If the medium fluctuations are assumed to be homogeneous, then the correlation function can be written as $B_n(\boldsymbol{\rho})$, a function of the difference coordinate,

$$\boldsymbol{\rho} = \mathbf{r}_1 - \mathbf{r}_2 \tag{10-74a}$$

For completeness in the transformation, the global coordinate [5] can be written

$$\boldsymbol{\eta} = (\mathbf{r}_1 + \mathbf{r}_2)/2 \tag{10-74b}$$

If the change in coordinates represented by Eqs. (10-74) is substituted in Eq. (10-73), the η integration can be performed directly, leaving a volume term outside,

$$\bar{I}(\mathbf{r}) = \frac{k^4 V A_0^2 \sin^2 \psi}{4\pi^2 r'^2} \iiint_V d\boldsymbol{\rho}\, B_n(\boldsymbol{\rho}) \exp\left[j\left(\mathbf{k} - \frac{k\mathbf{r}'}{r'}\right) \cdot \boldsymbol{\rho}\right] \tag{10-75}$$

Notice the characteristic dependence on the volume of the scatterer in the coefficient term.

If the volume of the scattering region is large compared to the region over which the correlation function has appreciable value, or if the correlation lengths are small compared to the characteristic linear dimension of the volume, then the integral in Eq. (10-75) represents a Fourier-transform operation, since the limits can be considered very large. If the appropriate factors of inverse 2π are incorporated, then the mean intensity can be written [141]

$$\bar{I}(\mathbf{r}') = \frac{2\pi k^4 V A_0^2 \sin^2 \psi}{r'^2} \Phi_n\left(\mathbf{k} - \frac{k\mathbf{r}'}{r'}\right) \tag{10-76}$$

where $\Phi_n(\mathbf{K})$ is a spectrum function for the medium fluctuations defined over the wave vector space by

$$\Phi_n(\mathbf{K}) = \frac{1}{8\pi^3} \iiint_{-\infty}^{\infty} d\boldsymbol{\rho}\, B_n(\boldsymbol{\rho}) e^{j\mathbf{K}\cdot\boldsymbol{\rho}} \tag{10-77}$$

and shown in Fig. 10-9. The associated inverse transform pair is†

$$B_n(\boldsymbol{\rho}) = \iiint_{-\infty}^{\infty} \Phi_n(\mathbf{K}) e^{-j\mathbf{K}\cdot\boldsymbol{\rho}}\, d\mathbf{K} \tag{10-78}$$

† Strictly speaking, one should start with Eq. (10-78), substitute in Eq. (10-75), and examine the inner integral over $\boldsymbol{\rho}$ of the exponent function. This leads to a function, whose limiting form is a delta function over **K** space.

The average or mean intensity given by Eq. (10-76) can be converted to an effective scattering cross section of the volume V by procedures similar to those used in the section on Mie–Debye scattering. The average intensity is multiplied by the $r^2\,d\Omega$, where $d\Omega$ is the differential solid angle, and then this product is divided by the incident amplitude squared $A_0{}^2$. The resulting differential cross section

$$d\sigma/d\Omega = 2\pi k^4 V \sin^2 \psi \Phi_n(\mathbf{k} - k\mathbf{r}'/r') \tag{10-79}$$

shows a marked dependence on all the parameters identified in the Rayleigh scattering case, except that one of the volume terms is replaced by $\Phi_n(\mathbf{K})$, which shows a marked peak at the origin of this wave vector space. Physically this means that a very narrow region of the turbulence spectrum contributes to the scattering at any particular scattering angle. Thus the amount of scattering is strongly determined by the character of the turbulence spectrum. The dependence on ψ retains the characteristic toroidal scattering associated with any dipole. That is, the maximum scattering occurs in directions orthogonal to the direction of the dipole.

There has been a great deal written [126, 139] about the various functional forms of the wave number spectrum of the refractive index fluctuations $\Phi_n(\mathbf{K})$. The principal results have been summarized by Strohbehn [126] and are basically twofold as shown in Fig. 10-9. Tatarski [142] has assumed, in the case that the temperature and velocity spectrums are the same, that the refractive index spectrum is

$$\Phi_m(K) = 0.033 C_n{}^2 K^{-11/3} \exp[-K^2/K_m{}^2] \tag{10-80}$$

where K_m times the inner scale of turbulence is 5.92, and $C_n{}^2$, the structure constant, is directly related to the strength of the refractive index fluctuations (typically 10^{-17} to 10^{-14} $\mathrm{m}^{-2/3}$). This representation has a problem with a singularity at K equal to zero, thus it does not describe the spectrum very well near the input region of the turbulence energy. This spectrum is also a modified form of the more classical inverse eleven-thirds Kolmogorov spectrum [5]. The modification arose primarily to handle the character of the spectrum in the dissipative region where no scattering spectrum should exist.

To handle the singularity problems near the origin of the wave number space, Strohbehn [120, 126] has used

$$\Phi_n(K) = 0.063 \langle n_1{}^2 \rangle L_0{}^3 \frac{\exp[-K^2/K_m{}^2]}{(1 + K^2 L_0{}^2)^{11/6}} \tag{10-81}$$

where $\langle \eta_1^2 \rangle$ is the variance of the refractive index fluctuations, and L_0 is the outer scale of turbulence in the input region. These various models representing the spectrum of atmospheric turbulence are still under development. There are problems with correspondence between theory and experiment for strong turbulence, as will be noted later.

The consistency of Eq. (10-76) with other results is, however, important. The inverse fourth-power wavelength dependence, the volume dependence, the dipole angular dependence, and the dependence on the spectrum of refractive index fluctuations are basic constituents in this representation. The theory that led to this representation is a first-order theory, and problems have been encountered in matching some of the measured moments with theory. In particular, experiments by Gracheva [126], Gurvich [126], and Deitz [143] have shown that this simple first-order theory does not account for an apparent saturation in the variance of intensity fluctuations as shown in Fig. 10-10.

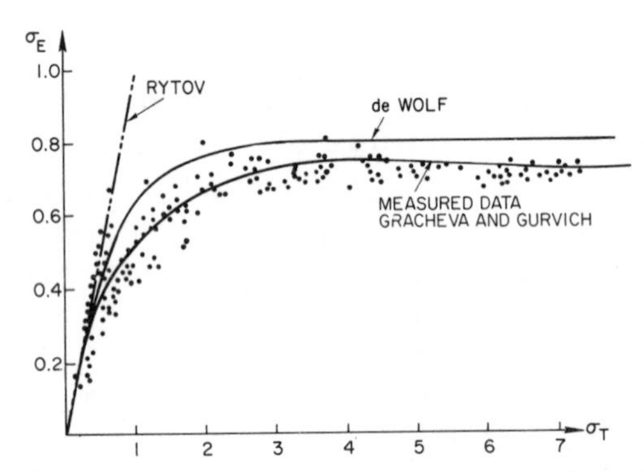

Fig. 10-10. Experimental values of $\sigma_{\ln I}$ plotted versus theoretical values of σ_T using corrected data of Graecheva and Gurvich (1965). The Rytov first-order result and deWolf theoretical models are plotted.

Attempts have been made to overcome the limitations of the weak perturbation method through the use of the method of smooth perturbations or Rytov's method [5]. This method, in essence, transforms the simple homogeneous wave equation into a Riccati equation by defining a function $\psi(\mathbf{r})$, using

$$E(\mathbf{r}) = e^{\psi(\mathbf{r})} \tag{10-82}$$

where

$$\psi(\mathbf{r}) = \chi(\mathbf{r}) + jS(\mathbf{r}) \tag{10-83}$$

Substitution of Eq. (10-82) into an equation like Eq. (10-69a) results in the Riccati equation

$$\nabla^2\psi + (\nabla\psi)^2 + k^2\eta^2 = 0 \tag{10-84}$$

If a perturbation parameter ξ is used to develop a perturbed expansion† of Eq. (10-84), then a hierarchy of equations result. This expansion is accomplished by representing the refractive index and $\psi(\mathbf{r})$ by

$$\eta(\mathbf{r}) = \langle\eta(\mathbf{r})\rangle + \xi\eta_1(\mathbf{r}) \tag{10-85a}$$

and

$$\psi(\mathbf{r}) = \psi_0(\mathbf{r}) + \xi\psi_1(\mathbf{r}) + \xi^2\psi_2(\mathbf{r}) + \sigma(\xi^3) \tag{10-85b}$$

where $\langle\eta(\mathbf{r})\rangle$ is approximately unity. The hierarchy of solutions obtained by substitution of Eqs. (10-85) into Eq. (10-84) and solving with Green's function techniques [144] is

$$\psi_1(\mathbf{r}) = \frac{k^2}{2\pi}\iiint_V d\mathbf{r}'\,\eta_1(\mathbf{r}')\frac{E_0(\mathbf{r}')}{E_0(\mathbf{r})}\,\frac{\exp[jk\,|\,\mathbf{r}-\mathbf{r}'\,|]}{|\,\mathbf{r}-\mathbf{r}'\,|} \tag{10-86a}$$

$$\psi_2(\mathbf{r}) = \frac{1}{4\pi}\iiint_V d\mathbf{r}'\{k^2\eta_1^2(\mathbf{r}') + \nabla\psi_1\cdot\nabla\psi_1\}\frac{E_0(\mathbf{r}')}{E_0(\mathbf{r})}\,\frac{\exp[jk\,|\,\mathbf{r}-\mathbf{r}'\,|]}{|\,\mathbf{r}-\mathbf{r}'\,|} \tag{10-86b}$$

$$\psi_n(\mathbf{r}) = \frac{1}{4\pi}\sum_{i=1}^{n-1}\iiint_V d\mathbf{r}'\{\nabla\psi_i\cdot\nabla\psi_{n-i}\}\frac{E_0(\mathbf{r}')}{E_0(\mathbf{r})}\,\frac{\exp[jk\,|\,\mathbf{r}-\mathbf{r}'\,|]}{|\,\mathbf{r}-\mathbf{r}'\,|} \tag{10-86c}$$

If a plane-wave assumption is made for $E_0(r)$ like that used in Eq. (10-72), then the following results for $\psi_1(\mathbf{r})$:

$$\begin{aligned}\psi_1(\mathbf{r}) &= \chi(\mathbf{r}) + jS_1(\mathbf{r})\\ &= \frac{k^2}{2\pi}\iiint_V d\mathbf{r}'\,\eta_1(\mathbf{r}')\exp[-j\mathbf{k}\cdot(\mathbf{r}-\mathbf{r}')]\frac{\exp[jk\,|\,\mathbf{r}-\mathbf{r}'\,|]}{|\,\mathbf{r}-\mathbf{r}'\,|}\end{aligned} \tag{10-87}$$

which can be considerably simplified in a small cone in the far field. However, the point of interest in experimental studies has been the variance of the amplitude fluctuations $\langle\chi^2\rangle$. Tatarski [5] obtained by spectral expansion techniques $\langle\chi^2\rangle$ for the plane-wave case as

$$\langle\chi^2\rangle = 0.31C_n^2k^{7/6}L^{11/6}, \qquad \sqrt{\lambda L} \gg l_0 \tag{10-88}$$

† The parameter ξ represents the smallness of the perturbation and enables a bookkeeping of like terms.

where L is the path length, and l_0 is the nominal inner scale size of the turbulence. Results for spherical waves [6], beam waves [145, 146], and waves in locally homogeneous media [10, 144] are similar except for magnification effects, focusing effects, and integrals over $C_n^2(\mathbf{r})$.

Since the intensity of the wave is given by

$$I = A_0^2 = e^{2\chi} \tag{10-89}$$

the variance of the log intensity fluctuations differs by a factor of four and is given by

$$\sigma_{\ln I} = \sigma_T = 1.23 C_n^2 k^{7/6} L^{11/6} \tag{10-90}$$

It is this intensity variance that is plotted in Fig. 10-10. These data are the Grachera and Gurvich (1965) data corrected by deWolf [147] and corroborated by the Deitz data [143]. The results show a dependence on range C_n^2 and wavelength as clarified by Clifford [148]. It is quite clear that the Rytov solution is not applicable for $\sigma_T > 0.3$. Thus above this value other effects are taking over, and the weak perturbation model is failing. A variance σ_I greater than unity has become a demarcation between the regions of weak turbulence and the regions of "strong turbulence." Fante [139] has given a review of all the related work surrounding the explanation of this saturation phenomena. Several authors [147, 149, 150] have developed an analytic method of describing this saturation effect using renormalization methods [142, 147] that involve selective summation techniques [151] and by modifications to the Rytov procedure, where more careful models for the irradiance statistics are used [152–154]. DeWolf has shown [147] that the intensity variance can be represented in the case of saturation by

$$\sigma^2 = \ln[2 - \exp(-\sigma_1^2)] \tag{10-91}$$

where σ_1^2 is given by Eq. (10-90). This curve has a functional form that "saturates" and is plotted as a solid line in Fig. 10-10. In essence, the methods account for the wave becoming incoherent as it propagates in the medium with various contributions from the incoherent scattering being appropriately canceled by other parts. Conceptually one should consider a partially coherent wave entering a thick phase medium, and the turbulent eddies are then scattering partially coherent waves. This concept is related to the work of Chapter IX.

Clifford [148] and Yura [155] have developed a plausible physical model for describing the process of saturation in strong turbulence. Their model is an extension of a model by Tatarski [142] in which the effect of scintillation is due to focusing at the receiver by the smallest turbulent eddies. In the new model, diffraction and the loss of transverse spatial coherence in the propagating wave have been added. The new model shows that the size of turbulence eddies most responsible for the scintillations is of the order of the coherence length and not $\sqrt{\lambda L}$. Clifford [148] has shown that intensity variance can be represented by a normalized covariance function

$$C_\chi(\varrho_n, \sigma_t^2) = 2.95\sigma_t^2 \int_0^1 du[u(1-u)]^{5/6} \int_0^\infty dy \frac{\sin^2 y}{y^{11/6}}$$
$$\times \exp[-\sigma_t^2[u(1-u)]^{5/6} f(y)] J_0\left\{\left(\frac{4\pi yu}{1-u}\right)^{1/2} \varrho_n\right\} \tag{10-92a}$$

where ϱ_n is a normalized transverse coordinate given by $\varrho/\sqrt{\lambda L}$, $f(y)$ is given by

$$f(y) = 7.02 y^{5/6} \int_{0.7y}^\infty d\xi\, \xi^{-8/3}[1 - J_0(\xi)] \tag{10-92b}$$

and σ_t is given by Eq. (10-90). Equation (10-92a) is the equation for the log amplitude variance and covariance function in the case of strong integrated turbulence ($\sigma_t^2 > 0.3$), and it reduces to the first-order result given by Eq. (10-90) for $\sigma_t^2 \leq 0.3$. This model and the results of Eqs. (10-92) do not saturate with the same flat response as shown by the deWolf model. Instead, the curves [148] show a slight negative slope in the saturation region, which seems to correspond with observed experimental data by Deitz [143] and Ochs and Lawrence [126, 148]. Hypothetically, the result of Eq. (10-92), if plotted in Fig. 10-10, would appear much as the solid curve sketched in the data points with a slight negative slope. This is not an explicit result since not all the data were available for a detailed comparison of these results. The Clifford result, however, shows good correspondence with their own data [148].

Extension of the results of this section to angle of arrival, phase fluctuation, temporal spectral effects, and many other system questions related to propagation of waves and beams in a turbulent atmosphere are discussed elsewhere [139]. In this section, the attempt has been to show the methodology used when the scatterers are random is shape and distribution. In addition, an attempt has been made to show the limitation of these theories.

Multiple Scattering in a System of Random Discrete Scatterers

So far the descriptions of scattering theories are basically within the context of single scattering. The specific effects of multiple scattering in a random media of discrete scatterers have been neglected. In the case of random continua, an explicit formulation for multiple scattering is more difficult to make. In essence, the techniques associated with selective summing [151] are a formalism that accounts for more than singular scattering events. In the case of a random distribution of discrete scatterers, the formulation is simpler in at least the conceptual sense. Historically, the problem of scattering by random discrete scatterers has been investigated by many [156–158] from the phenomenological and the analytical points of view. The phenomenological approach is associated with transport theory [159–161] or radiative transfer theory [131]. These theories deal with propagation of specific intensities or brightness and have been extensively used in geophysical problems. The methods, however, do not deal with field fluctuations or correlation functions. The analytical theory works directly with field quantities and their associated intensities and correlation functions. The diagram method described by Frisch [151] and the multiple scattering formalism of Twersky [47, 158] are examples of the analytical method. Recent papers [128–130, 142, 159, 160, 162, 163] have described the correspondence between the two approaches to multiple scattering and have led to an increased understanding of the multiple scattering processes. One example is the explanation of pulse broadening and pulse tails in fog [129] which had recently been measured [164].

Since all the scattering theories are dealing with the same basic physical phenomena, it would seem from all the different descriptions of scattering processes that the understanding of light scattering is fragmented. This is not true, particularly as a result of many recent publications linking these theories. It is, however, important to note that an adequate understanding of these very complex theories can best be attained by a partitioning of the theories into domains associated with their applicability. It would seem that someone would soon prepare a tome that would link the theories and experiments into a very understandable composite [165]. Until then it is necessary to examine these ideas in a somewhat piecemeal fashion.

The Twersky multiple scattering theory, for a simple case, is a reasonable place to begin [13]. In essence, the method involves the use of scattering functions for each scatterer and decomposition into coherent and incoherent fields and intensities. Integral equations describing these phenomena are derived for the coherent (average) field and intensities.

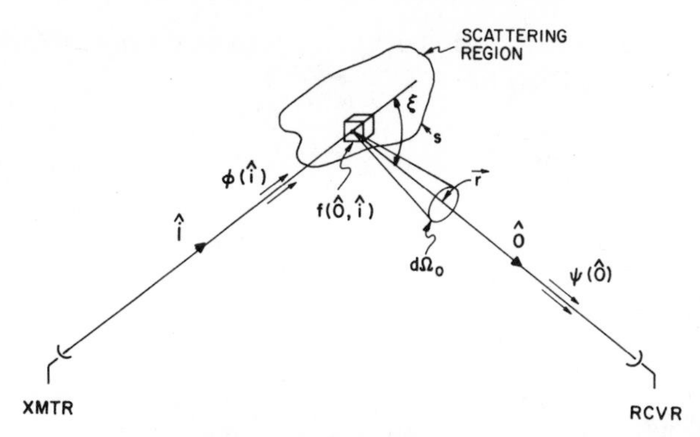

Fig. 10-11. Coordinate framework for multiple scattering in a region of random discrete scatterers having scattering amplitudes $f(\hat{o}, \hat{i})$.

To illustrate the method, a single scatterer is first considered in this formalism. An incident plane wave $\phi(\hat{i})$ is assumed in a system, where the incident wave vectors are aligned with the incident unit vector $\hat{i}$, and the scattered wave u is aligned with $\hat{o}$. The total wave at the receiver is $\psi(\hat{o}, \hat{i})$, and it is composed of the primary wave component ϕ and the scattered wave component u. The system is illustrated in Fig. 10-11. The scattering results from a small isotropic scatterer, and the angular character of this scattering is given by a scattering amplitude function $f(\hat{o}, \hat{i})$, which is independent of r. The radial coordinate r is centered in the smallest sphere enclosing the scatterer. In general, it is assumed that this scattering amplitude is not affected by adjacent scatterers and that a system of multiple scatterers reduces to a single scattering equivalent as the other scatterers recede to infinity. Thus the waves involved in this single system can be represented by [131]

$$\phi(\hat{i}) = e^{j\mathbf{k}\cdot\mathbf{r}}, \qquad \mathbf{k} = k\hat{i}, \qquad \mathbf{r} = r\hat{o} \tag{10-93}$$

and

$$\psi - \phi = u \sim f(\hat{o}, \hat{i})(e^{jkr}/r) \tag{10-94}$$

where the total scattered field ψ satisfies the wave equation with appropriate boundary conditions, and the boundary condition on u is the radiation condition.†

Imposing the lossless constraint on the scatterer, represented by [137]

$$\iint_S \left[\psi^*(\hat{i})\,\frac{\partial\psi(\hat{j})}{\partial n} - \psi(\hat{j})\,\frac{\partial\psi^*(\hat{i})}{\partial n}\right] dS = 0 \tag{10-95}$$

† See Chapter I for a discussion of the radiation condition [137] and Green's theorem.

gives a formalism for analytically representing $u(\hat{o})$. If the function $E(k \mid \mathbf{r} - \mathbf{r}' \mid)$ is defined as

$$E(k \mid \mathbf{r} - \mathbf{r}' \mid) \equiv \frac{\exp(jk \mid \mathbf{r} - \mathbf{r}' \mid)}{\mid \mathbf{r} - \mathbf{r}' \mid} \tag{10-96}$$

where $\mathbf{r}'$ represents the vector position of the receiver, then Green's theorem† and Eq. (10-96), in a region outside the scatterer, lead to a representation for $u(\mathbf{r})$

$$u(\mathbf{r}) = -\frac{1}{4\pi} \iint_S \left[E(k \mid \mathbf{r} - \mathbf{r}' \mid) \frac{\partial u(\mathbf{r}')}{\partial n} - u(\mathbf{r}') \frac{\partial E}{\partial n} \right] dS(\mathbf{r}') \tag{10-97}$$

This expression can be used on a very large boundary sphere to show [94] that

$$f(\hat{o}, \hat{\imath}) = -\frac{1}{4\pi} \iint_S \left[\exp(-jk\hat{o} \cdot \mathbf{r}') \frac{\partial \psi(\mathbf{r}', \hat{\imath})}{\partial n} - \psi(\mathbf{r}', \hat{\imath}) \frac{\partial \exp(-jk\hat{o} \cdot \mathbf{r}')}{\partial n} \right] dS \tag{10-98}$$

which leads to the following properties† of $f(\hat{o}, \hat{\imath})$

$$f(\hat{\imath}, \hat{\jmath}) = f(-\hat{\jmath}, \hat{\imath}) \tag{10-99a}$$

$$(-2\pi j/k)[f(\hat{\jmath}, \hat{\imath}) - f^*(\hat{\imath}, \hat{\jmath})] = \iint_\pi f^*(\hat{o}, \hat{\jmath}) f(\hat{o}, \hat{\imath})\, d\Omega_o \tag{10-99b}$$

and

$$(4\pi/k) \operatorname{Im} f(\hat{\imath}, \hat{\imath})\} = \iint_\pi \mid f^*(\hat{o}, \hat{\imath}) \mid^2 d\Omega_o = \sigma_s(\hat{\imath}) \tag{10-99c}$$

where σ_s is the total scattering cross section, and Ω_o is the solid angle about $\hat{o}$. The scattering amplitude is then related to the notion of scattering cross section that was presented in earlier sections and provides a means for physically understanding its origin. It is in general complex, implying that in some sense waves attenuate in a multiple scattering medium.

For a system in which the scatterer is located at $\mathbf{r}_s$, the observer at $\mathbf{r}_a$, the excitation is $\exp[j\mathbf{k} \cdot (\mathbf{r} - \mathbf{r}_s')]$, and $\mathbf{r}'$ is a vector from $\mathbf{r}_s$ to a point on an enclosing surface $S(\mathbf{r}')$, the scattered wave $u_s^a(\hat{\imath})$ can be represented by

$$u_s^a(\hat{\imath}) = \frac{jk}{2\pi} \iint_S \exp[jk\mathbf{p} \cdot (\mathbf{r}_a - \mathbf{r}_s)] f(\hat{p}, \hat{\imath})\, d\Omega_p \tag{10-100}$$

† It is an unfortunate notation that uses $\hat{\jmath}$ for a unit vector and j for $\sqrt{-1}$. This is done to be consistent with the literature.

where $\iint d\Omega_p$ is a Sommerfeld [15] kind of complex double integral. In general, Eq. (10-100) can be solved by a saddle-point approximation [66] or represented by a convergent series in inverse powers of r and angular derivatives of $f(\hat{o}, \hat{\imath})$ like

$$u_s{}^a(\hat{\imath}) = u_s(\mathbf{r}_a - \mathbf{r}_s; \hat{\imath}) = \frac{e^{jkr}}{r}\left\{1 + \frac{j}{2kr} D + \left(\frac{j}{2kr}\right)^2 \frac{D(D-2)}{2!} + \cdots + \left(\frac{1}{n!}\right)\left(\frac{j}{2kr}\right)^n \left[D(D-2!) \cdots \left(D - \frac{n!}{(n-2)!}\right)\right]\right\} \times f_s\left(\frac{\mathbf{r}_a - \mathbf{r}_s}{|\mathbf{r}_a - \mathbf{r}_s|}, \hat{\imath}\right) \tag{10-101a}$$

where the angular derivative operator D is

$$D = \left[-\frac{1}{\sin^2\theta}\right]\left[\frac{\partial^2}{\partial\phi^2} + \sin\theta \frac{\partial}{\partial\theta}\left(\sin\theta \frac{\partial}{\partial\theta}\right)\right] \tag{10-101b}$$

This detail has been given to illustrate the operator relationship of $\phi(\hat{\imath})$, $E(\ )$, and $f(\hat{o}, \hat{\imath})$. This operator formalism becomes much more important when discussing multiple scatterers. Notice that in the scattered waves representation of Eqs. (10-100) and (10-101), the exciting field $\phi_s(\hat{\imath})$ is not an explicit part of the representation. To incorporate the primary field and thus yield a scattered field at $\mathbf{r}_a$, the following is used:

$$u_s{}^a\phi_s = u_s{}^a(\hat{\imath})\phi_s(\hat{\imath}) = E_{as}f_s\phi_s \tag{10-102a}$$

and

$$\phi_s = \exp[jk\hat{\imath} \cdot \mathbf{r}_s] \tag{10-102b}$$

where Eq. (10-100) is implicit in E_{as}. The notation of Eq. (10-102a) is important for understanding the notation of the multiple scattering situation.

For many scatterers, N, not necessarily identical and located at $\mathbf{r}_N$, the solution outside the scatterers' surface is

$$\psi(\mathbf{r}) = \phi + U = \phi + \sum_{s=1}^{N} U_s(\mathbf{r} - \mathbf{r}_s) \tag{10-103a}$$

where by extension

$$U_s^a = U_s(\mathbf{r}_a - \mathbf{r}_s) = E_{as}F_s \tag{10-103b}$$

The term U_s is the multiple scattered wave, and F_s is the multiple scattered amplitude, which reduce to their single scattered analogs as $|\mathbf{r}_s - \mathbf{r}_t|$

goes to infinity for fixed $| \mathbf{r}_s - \mathbf{r}_a |$,

$$U_s^a \to u_s^a \phi_s \tag{10-104a}$$

and

$$F_s \to f_s \phi_s \tag{10-104b}$$

The difference now for the multiple scattering case is that the exciting field Φ_s for scatterer s is composed of the primary wave $\phi_s(\hat{i})$ plus the scattered waves from all other scatterers

$$\Phi_s = \phi_s + \sum_{t,t\neq s}^{N} u_t^s \Phi_t \tag{10-105}$$

where superposition and the form of Eqs. (10-104) has been used. Thus one can use Eq. (10-103) and Eq. (10-105) to write a compact notation for the total scattered field as

$$\psi_c = \phi_c + \sum_s E_{cs} F_s = \phi_s + \sum E_{cs} f_s \Phi_s \tag{10-106a}$$

and

$$F_s = f_s \Phi_s = f_s \phi_s + \sum_{t,t\neq s} E_{st} f_s F_t \tag{10-106b}$$

which, in terms of the plane-wave result represented by Eq. (10-98), gives a simpler set of integral equations

$$\psi_c = \exp(jk\hat{i} \cdot \mathbf{r}_c) + \frac{jk}{2\pi} \sum_s \iint \exp[jk\hat{p} \cdot (\mathbf{r}_c - \mathbf{r}_s)] F_s(\hat{p}, \hat{i})\, d\Omega_p \tag{10-107a}$$

and

$$F_s(\hat{o}, \hat{i}) = f_s(\hat{o}, \hat{i}) \exp(jk\hat{i} \cdot \mathbf{r}_s) + \frac{jk}{2\pi} \sum_{t,t\neq s} \iint f_s(\hat{o}, \hat{p}) \exp[jk\mathbf{p} \cdot (\mathbf{r}_s - \mathbf{r}_t)] \times F_t(\hat{p}, \hat{i})\, d\Omega_0 \tag{10-107b}$$

Equations (10-107) represent a solution in a random variable sense for the N particles. This type of solution is not very useful in an explicit sense. What is useful are the various moments (means, correlations, etc.) of the random functions. To evaluate these moments, the probability distribution functions are necessary [137]. In general, these are obtained by making plausibility arguments concerning the nature of the system and the constraining physics. However, for the sake of this description, a simple assumption will be made. The scatterers will be assumed to be independent of each other and uniformly distributed. This permits the use of a number density, $\varrho(s)$, which is N divided by the volume. Using this formalism, one can write the average or coherent scattered field as

$$\langle \psi_c \rangle = \phi_c + \iiint \varrho(\mathbf{s}) u_s{}^c \langle \Phi_s \rangle \, d\mathbf{s} \tag{10-108a}$$

with

$$\langle \Phi_s \rangle = \phi_s + \sum_{t, t \neq s} \iiint u_t{}^s \langle \Phi_t \rangle_{st} P_{st}(\hat{s}, \hat{t}) \, d\hat{t} \tag{10-108b}$$

where $P_{st}(\hat{s}, \hat{t})$ is a conditional probability [137] on a two-particle scattering event, and $d\mathbf{s}$ is a volume integral. In general, reduction of Eq. (10-108b) leads to a hierarchy of equations in which one considers the probability of three scatter events, four scatter events, and so on until an N scatter event is considered which, in essence, gives a general multiple scattering event sequence. Although Eq. (10-108b) is important for completeness, one generally deals only with Eq. (10-108a) and makes approximations about $\langle \Phi_s \rangle$. In this regard Foldy [156] assumed that $\langle \Phi_s \rangle$ could be replaced by $\langle \psi_s \rangle$, the exciting field without scatterer s present, thus yielding a closed integral equation for which solutions were sought. This simple form is good for dilute systems in which an individual scatterer has little effect on adjacent scatterers. Lax [157] introduced the approximation of a field factor c, which modified the assumption to $c\langle \psi_s \rangle$, where c depends on the correlation between pairs of scatterers. In general then, it is reasonable [128] for most dilute systems to write the coherent scattered field as

$$\langle \psi_c \rangle = \phi_s + \iiint u_s{}^c \langle \psi_s \rangle \varrho(\mathbf{s}) \, d\mathbf{s} \tag{10-109}$$

where $u_s{}^c$ is an operator in the sense of Eq. (10-101a).

Using the first term of Eq. (10-101a) in the far zone of the scatterer, an expression for $u_s{}^c \langle \psi_s \rangle$ can be given. One recognizes, however, that this is basically fE_{as}, where E_{as} is the free-space Green's function of a monopole radiator. Thus the application of $(\nabla^2 + k^2)$ to both sides of Eq. (10-109) results in a scattered wave that behaves locally like a plane wave satisfying a modified wave equation

$$(\nabla^2 + K^2)\langle \psi(\mathbf{r}) \rangle = 0 \tag{10-110}$$

where

$$K^2 = k^2 + (2\pi/k) f(\hat{i}, \hat{i}) \varrho \tag{10-111}$$

This result follows from the recognition that $(\nabla^2 + k^2)$ operating on a monopole Green's function yields a delta function in the integrand of

Eq. (10-109). The delta function reduces the integral to the second term in Eq. (10-111). This result is also important in Eq. (10-99c), since it can be expanded to include absorption effects σ_a in the total cross section σ_t

$$(4\pi/k)\,\mathrm{Im}\{f(\hat{i}, \hat{i})\} = \sigma_s + \sigma_a = \sigma_t \tag{10-112}$$

The absorption effect arises because K is, in general, complex

$$K = K_r + jK_i = K_r + j(\varrho\sigma_t/2) \tag{10-113}$$

This implies that the coherent intensity $|\langle\psi\rangle|^2$ attenuates as $\exp[-\varrho\sigma_t r]$. This result is important in that now the multiple scattering analysis is equivalent to an attenuating coherent wave propagating in an equivalent media with a reduced phase velocity. That is, the coherent wave slows down and attenuates.

The coherent intensity was mentioned in the previous section. This concept is completed in an energy sense by defining the coherent intensity in relation to the total intensity $\langle|\psi|^2\rangle$, which is the average of the square of the magnitude of the total field, yielding [137, 156]

$$\langle|\psi_c|^2\rangle = |\langle\psi_c\rangle|^2 + \langle|\psi_i|^2\rangle \tag{10-114}$$

Equation (10-114) basically becomes a defining relation for the second term, which is called the incoherent intensity. This term can be expressed in terms of the scattered wave U_s as [137]

$$\langle|\psi_i|^2\rangle = \iiint \varrho(\mathbf{s})\,|\langle U_s\rangle|^2\,d\mathbf{s} \tag{10-115}$$

where Eq. (10-103b) must be used in reductions of Eq. (10-115).

An integral equation [128, 137] can be derived for the coherent intensity by a simplification of the integral equation for the correlation function of the coherent field [128]

$$\langle\psi_a\psi_b^*\rangle = \langle\psi_a\rangle\langle\psi_b^*\rangle + \iiint v_s^a v_s^{b*}\langle|\psi_s|^2\rangle\varrho(\mathbf{s})\,d\mathbf{s} \tag{10-116}$$

where v_s^a is like u_s^a only in the new equivalent media. That is,

$$v_s^a = u_s^a + \iiint u_t^a v_s^t \varrho(\mathbf{t})\,d\mathbf{t} \tag{10-117}$$

where in an approximate sense

$$v_s^a = f(\hat{o}, \hat{i})\,\frac{\exp[jK\,|\mathbf{r}_a - \mathbf{r}_s|]}{|\mathbf{r}_a - \mathbf{r}_s|} \tag{10-118}$$

and K is defined by Eq. (10-111). The coherent intensity integral equation is then

$$\langle | \psi_a |^2 \rangle = | \langle \psi_a \rangle |^2 + \iiint v_s^{a} v_s^{a*} \langle | \psi_s |^2 \rangle \varrho(\mathbf{s}) \, d\mathbf{s} \qquad (10\text{-}119)$$

Equation (10-119) completes the major result of the multiple scattering theory as derived by Twersky [156].

Ishimaru [128, 129] has used this formulation, specifically Eqs. (10-116) and (10-117), to derive the radiative transfer equation [161] for the specific intensity $I(\mathbf{r}, \hat{\imath})$. This follows by defining

$$\langle \psi_a \psi_b^* \rangle = \Gamma(\mathbf{r}_a, \mathbf{r}_b) \simeq \iint I(\mathbf{r}, \hat{\imath}) \exp[jK_r \hat{\imath} \cdot (\mathbf{r}_a - \mathbf{r}_b)] \, d\Omega \qquad (10\text{-}120)$$

which correspond with the definitions of mutual coherence functions given in Chapter IX. Also $\langle \psi^a \rangle \langle \psi^{b*} \rangle$ is defined in terms of the specific coherent intensity by an equation like Eq. (10-119) except that $I(\ )$ goes to $I_c(\ )$. It now follows [128], with approximations in the far zone of the scatterer, that

$$(d/ds) I(\mathbf{r}, \hat{s}) = -\varrho \sigma_t I(\mathbf{r}, \hat{s}) + \iint \varrho(s) \, | f(\hat{s}, \hat{s}') |^2 \, I(\mathbf{r}, \hat{s}') \, d\Omega' \qquad (10\text{-}121)$$

which is the radiative transfer equation [161]. It is implicit in the derivation of Eq. (10-121) that $I_c(\mathbf{r}, \hat{s})$ satisfies a simple rate equation

$$(d/ds) I_c(\mathbf{r}, \hat{s}) = -\varrho \sigma_t I_c(\mathbf{r}, \hat{s}) \qquad (10\text{-}122)$$

The solutions of Eq. (10-122) proves the assertion made about attenuation of the propagating coherent intensity. This coupling between the phenomenological and analytical approaches to multiple scattering is very important to the establishment of a basic scattering theory.

One can go even further to show that the mutual coherence function used in turbulence theory satisfies a parabolic equation [142, 159, 160] very similar to Eq. (10-121). Fante [159] started with the basic transport equation [161] to show such a correspondence. Extensions of these results [129, 160] have led to results that correspond with experiment [164]. The pulse broadening in dense media is such an example [129]. These correspondences between the many scattering theories have provided a basis for coalescing of the understanding of general scattering processes. The mathematics is still very difficult, but the conceptual framework is now quite clear [165].

Summary

In this chapter an overview of scattering theories has been presented. The presentation began with a heuristic approach to explaining physical phenomena that are readily observable—skylight, rainbows, polarization, and others. With this intuitive notion established, a more detailed effort was given explaining discrete, deterministic, dielectric particle scattering, and absorption effects. The simple explanation of the Rayleigh scattering and its inverse fourth-power wavelength dependence for small, dilute particles was contrasted with the more detailed Mie scattering for large spherical particles. The work then dealt with the dilemma imposed by nondeterministic continua and particles. The random media scattering development explored the first-order theories and showed the difficulty in handling the strong turbulence situation and its associated scintillation variance saturation. The final section provided a sketch of how to handle multiple scattering by discrete random scatterers. The formalism of this method is ponderous, but the recent coupling to other theories has enabled a broader understanding and coalescing of the phenomenological and analytical theories. The relationship of scattering theory to partial coherence theory also provided another link into current research work. It would seem that many of these methods will be simplified and that the correspondences with experiment will demonstrate a clearer understanding of some very interesting physical phenomena.

Problems

1. Calculate the ratio of scattered intensity of blue (4500 Å) skylight over red skylight (6300 Å).

2. Trace out the path of rays in the water droplet for the secondary rainbow.

3. Find the change in magnitude of the internal field of a dielectric sphere in air if ε_r for the sphere is assumed to be 9.

4. Find the magnitude of the dipole moment $|\mathbf{p}|$ for the sphere described in Problem 3 if the radius is assumed to be 1 mm.

5. If the dipole moment of a sphere is assumed to lie along the x direction in an (x, y, z) coordinate system, derive an expression for $\sin\psi$ in terms of spherical coordinate variables ϱ, θ, and ϕ.

6. Show how Eq. (10-66) follows from Eq. (10-64).

7. Show how Eq. (10-67) follows from Eq. (10-66) using the perturbed refractive index model, $\eta = 1 + \eta_1$.

8. Show that Eq. (10-71) does indeed reduce to a transverse field, Eq. (10-73), in a Fraunhofer zone.

9. Demonstrate that Eq. (10-76) can be obtained by starting with Eq. (10-78) and substituting in Eq. (10-75).

10. Show how Eq. (10-84) follows from the assumption of Eq. (10-82).

11. Show how Eqs. (10-86) follow from Eq. (10-84) using Eqs. (10-85).

12. Using the conditions of reciprocity and the conditions of Eqs. (10-98), (10-95), and (10-94), derive Eqs. (10-99).

13. Expand the third and fourth terms of Eq. (100-101a) using Eq. (10-101b).

14. Using the inhomogeneous wave equation $(\nabla^2 + k^2)G = -\delta(r - r')$, show that Eq. (10-111) results.

15. Show that the simple attenuation equation $\exp[-\varrho\sigma_t r]$ that follows Eq. (10-113) can be derived from Eq. (10-122).

REFERENCES

[1] S. Ramo, J. R. Whinnery, and T. VanDuzer, "Fields and Waves in Communication Electronics." Wiley, New York, 1965.

[2] J. A. Stratton, "Electromagnetic Theory." McGraw-Hill, New York, 1941.

[3] M. Kline and I. W. Kay, "Electromagnetic Theory and Geometrical Optics." Wyley (Interscience), New York, 1965.

[4] M. Kerker, "The Scattering of Light and Other Electromagnetic Radiation." Academic Press, New York, 1969.

[5] V. I. Tatarski, "Wave Propagation in a Turbulent Medium." McGraw-Hill, New York, 1961.

[6] G. Goubau and F. Schwering, On the guided propagation of electromagnetic beams, *IRE Trans. Antennas Propagation* **AP-9**, 248 (1961).

[7] J. R. Pierce, Modes in sequences of lenses, *Proc. Nat. Acad. Sci. U.S.* **47**, 1808 (1961).

[8] L. Levi, "Applied Optics, A Guide to Optical System Design." Wiley, New York, 1968.

[9] N. S. Kapany and J. J. Burke, "Optical Waveguides." Academic Press, New York, 1972.

[10] F. P. Carlson, Application of optical scintillation measurements to turbulence diagnostics, *J. Opt. Soc. Am.* **59**, 1343 (1969).

[11] M. Born and E. Wolf, "Principles of Optics," 5th ed. Pergamon, Oxford, 1975.

[12] L. I. Schiff, "Quantum Mechanics." McGraw-Hill, New York, 1955.

[13] J. W. Goodman, "Introduction to Fourier Optics." McGraw-Hill, New York, 1968.

[14] H. W. Reddick and F. H. Miller, "Advanced Mathematics for Engineers," 3rd ed. Wiley, New York, 1960.

[15] A. Sommerfeld, "Optics," Vol. IV. Academic Press, New York, 1964.

[16] E. Jahnke, F. Ende, and F. Lösch, "Tables of Higher Functions," 6th ed. McGraw-Hill, New York, 1960.

[17] K. Knopp, "Theory and Application of Infinite Series." Hafner, New York, 1947.

[18] J. M. Stone, "Radiation and Optics, An Introduction to the Classical Theory." McGraw-Hill, New York, 1963.

[19] D. Halliday and R. Resnick, "Physics." Wiley, New York, 1963.

[20] C. S. Williams and O. A. Becklund, "Optics, A Short Course for Engineers and Scientists." Wiley (Interscience), New York, 1972.

[21] J. B. Thomas, "An Introduction to Statistical Commutation Theory." Wiley, New York, 1969.

[22] L. J. Cutrona, E. N. Leith, C. J. Palermo, and L. J. Porcello, Optical data processing and filtering systems, *IRE Trans. Inform. Theory* **IT-6**, 386 (1960).

[23] E. L. O'Neill, Spatial filtering in optics, *IRE Trans. Inform. Theory* **IT-2**, 56 (1956).

[24] J. T. Tippett *et al.* (eds.), "Optical and Electro-optical Information Processing." MIT Press, Cambridge, Massachusetts, 1965.

[25] A. Papoulis, "Systems and Transforms with Applications in Optics." McGraw-Hill, New York, 1968.

[26] R. Courant and D. Hilbert, "Methods of Mathematical Physics," Vol. I. Wiley (Interscience), New York, 1953.

[27] E. L. O'Neill, "Introduction to Statistical Optics." Addison-Wesley, Reading, Massachusetts, 1963.

[28] W. T. Cathey, "Optical Information Processing and Holography." Wiley, New York, 1974.

[29] G. R. Fowles, "Introduction to Modern Optics." Holt, New York, 1968.

[30] E. H. Linfoot, "Fourier Methods in Optical Image Evaluation." Focal Press, London, 1964.

[31] A. Vander Lugt, Operational notation for the analysis and synthesis of optical data-processing systems, *Proc. IEEE* **54**, 1055 (1966).

[32] G. W. Stroke, "An Introduction to Coherent Optics and Holography." Academic Press, New York, 1966.

[33] A. R. Shulman, "Optical Data Processing." Wiley, New York, 1970.

[34] E. Abbe, *Arch. Mikroskop. Anal.* **9**, 413 (1873).

[35] A. B. Porter, On the diffraction theory of microscope vision, *Phil. Mag.* (6), **11**, 154 (1906).

[36] F. P. Carlson and R. Eguchi, Linear vector operations in coherent optical data processing systems, *Appl. Opt.* **9**, 687 (1970).

[37] A. Maréchal and M. Francon, "Diffraction," Editions de la Revue d'Optique, Paris, 1960.

[38] G. L. Turin, An introduction to matched filters, *IRE Trans. Inform. Theory* **IT-6**, 311 (1960).

[39] A. Vander Lugt, Signal detection by complex spatial filtering, *IRE Trans. Inform. Theory* **IT-10**, 139 (1964).

[40] J. E. Ward III, F. P. Carlson, and J. D. Heywood, Coherent optical recognition and counting of reticulated red blood cells, *IEEE Trans. Bio. Med. Eng.* **BME-21**, 12 (1974).

[41] K. Preston, Jr., "Coherent Optical Computers." McGraw-Hill, New York, 1972.

[42] L. J. Cutrona, E. N. Leith, L. J. Porcello, and W. E. Vivian, On the application of coherent optical processing techniques to synthetic-aperture radar, *Proc. IEEE* **54**, 1026 (1966).

[43] L. C. Graham, Synthetic interferometer radar for topographic mapping, *Proc. IEEE* **62**, 763 (1974).

[44] J. W. Goodman, Synthetic-aperture optics, *Progr. Opt.* **8**, 3 (1970).

[45] Kodak Plates and Films for Scientific Photography, P-315. Eastman Kodak Co., 1973.

[46] G. C. Higgins and L. A. Jones, Evaluation of image sharpness, *J. Soc. Motion Picture Television Eng.* **58**, 277 (1952).
[47] T. H. James and G. C. Higgins, "Fundamentals of Photographic Theory." Morgan and Morgan, 1960.
[48] Modulation transfer function, *Seminar Proc. SPIE* **13**. Boston, Massachusetts, 1969.
[49] M. Ross, "Laser Receivers, Devices, Techniques and Systems." Wiley, New York, 1966.
[50] K. Bromley, An optical incoherent correlator, *Opt. Acta* **21**, 35 (1974).
[51] R. P. Bocker, Matrix multiplication using incoherent optical techniques, *Appl. Opt.* **13**, 1670 (1974).
[52] I. Schneider, M. Marrone, and M. Kabler, Dichroic absorption of M centers: A basis for optical information storage, *Appl. Opt.* **9**, 1163 (1970).
[53] F. Lüty, "Physics of Color Centers." Academic Press, New York, 1968.
[54] C. H. Holbrow and W. C. Davidon, An introduction to dispersion relations, *Am. J. Phys.* **32**, 762 (1964).
[55] M. Sharnoff, Validity conditions for Kramer-Kronig relations, *Am. J. Phys.* **32**, 40 (1964).
[56] I. Schneider, Dispersion and diffraction by anisotropic centers in alkali-halides, *Phys. Rev. Lett.*, **32**, 412 (1974).
[57] D. Marcuse, "Theory of Dielectric Optical Waveguides." Academic Press, New York, 1974.
[58] L. F. Mollenauer and D. H. Olson, A broadly tunable cw laser using color centers, *Appl. Phys. Lett.* **24**, 386 (1974).
[59] M. Beran and G. Parrent, Jr., "Theory of Partial Coherence." Prentice Hall, Englewood Cliffs, New Jersey, 1964.
[60] P. H. Deitz and F. P. Carlson, Intensity interferometry in the spatial domain, *J. Opt. Soc. Am.* **63**, 274 (1973).
[61] P. H. Deitz and F. P. Carlson, Spatial irradiance interferometry with sources of arbitrary symmetry, *J. Opt. Soc. Am.* **64**, 11 (1974).
[62] D. Gezari, A. Labeyrie, and R. Stachnik, Speckle interferometry: Diffraction limited measurement of nine stars with the 200 inch telescope, *Astrophys. J.* **173**, L1-L5 (1972).
[63] L. A. Weinstein, "Open Resonators and Open Waveguides." Golem Press, Boulder, Colorado, 1969.
[64] A. G. Fox and T. Li, Resonant modes in a maser interferometer, *BSTJ* **40**, 453 (1961).
[65] G. P. Boyd and J. P. Gordon, Confocal multimode resonator for millimeter through optical wavelength masers, *BSTJ* **40**, 489 (1961).
[66] P. M. Morse and H. Feshbach, "Methods of Theoretical Physics," Part I. McGraw-Hill, New York, 1953.
[67] G. P. Boyd and H. Kogelnik, Generalized confocal resonator theory, *BSTJ* **41**, 1347 (1962).
[68] D. Gabor, A new microscopic principle, *Nature* (*London*) **161**, 777 (1948).
[69] D. Gabor, Microscopy by reconstructed wavefronts, *Proc. Roy. Soc.* **A197**, 454 (1949).
[70] D. Gabor, Microscopy by reconstructed wavefronts, 11, *Proc. Phys. Soc.* **B64**, 449 (1951).

[71] E. N. Leith and J. Upatnieks, Reconstructed wavefronts and communication theory, *J. Opt. Soc. Am.* **52**, 1123 (1962).
[72] E. N. Leith and J. Upatnieks, Wavefront reconstruction with continuous-tone objects, *J. Opt. Soc. Am.* **53**, 1377 (1963).
[73] E. N. Leith and J. Upatnieks, Wavefront reconstruction with diffused illumination and three-dimensional objects, *J. Opt. Soc. Am.* **54**, 1295 (1964).
[74] R. J. Collier, C. B. Burckhardt, and L. H. Lin, "Optical Holography." Academic Press, New York, 1971.
[75] J. B. DeVelis and G. O. Reynolds, "Theory and Applications of Holography." Addison-Wesley, Reading, Massachusetts, 1967.
[76] A. F. Metherell, H. M. El-Sum, and L. Larmore, "Acoustical Holography," Vol. I. Plenum Press, New York, 1969.
[77] A. F. Metherell and L. Larmore, "Acoustical Holography," Vol. II. Plenum Press, New York, 1970.
[78] J. Upatnieks, A. Vander Lugt, and E. N. Leith, Correction of lens aberrations by means of holograms, *Appl. Opt.* **5**, 589 (1966).
[79] "Modern Optics" (*Proc. Symp. Vol. XVII*). Polytechnic Press, New York, 1967.
[80] D. Gabor, Character recognition by holography, *Nature* (*London*) **208**, 422 (1965).
[81] K. A. Haines and B. P. Hildebrand, Interferometric measurement on diffuse surfaces by holographic techniques, *IEEE Trans. Instrum. Measurements* **IM-15**, 149 (1966).
[82] A. Kozma, "Introduction to Optical Data Processing." McGraw-Hill, New York, 1967.
[83] K. A. Haines and B. P. Hildebrand, Surface-deformation measurement using the wavefront reconstruction technique, *Appl. Opt.* **5**, 595 (1966).
[84] R. L. Powell and K. A. Stetson, Interferometric vibration analysis by wavefront reconstruction, *J. Opt. Soc. Am.* **55**, 1593 (1965).
[85] B. P. Hildebrand, General theory of holography, *J. Opt. Soc. Am.* **60**, 1511 (1970).
[86] A. Richter and F. P. Carlson, Holographically generated lens, *Appl. Opt.* **13**, 2924 (1974).
[87] J. E. Ward III, F. P. Carlson, and D. C. Auth, Lens aberration correction by holography, *Appl. Opt.* **10**, 896 (1971).
[88] F. P. Carlson, Generalized optical operators, *IEEE Intercon. Conv. Record* **7**, 1973.
[89] D. Gabor, Holography or the 'Whole Picture,' *New Scientist* 74 (1966).
[90] L. H. Lin, K. S. Pennington, G. S. Stroke, and A. E. Labeyrie, Multicolor holographic image reconstruction with white light illumination, *BSTJ* XIV, **4**, 659 (1966).
[91] E. Verdet, Leçons d'optique physique, *Paris l'Imprimerie Impériale* **1**, 106 (1869).
[92] L. Mandel and E. Wolf, "Selected Papers on Coherence and Fluctuations of Light," Vol. I. Dover, New York, 1970.
[93] L. Mandel and E. Wolf, "Selected Papers on Coherence and Fluctuations of Light," Vol. II. Dover, New York, 1970.
[94] F. Zernike, The concept of degree of coherence and its application to optical problems, *Physica* **5**, 785 (1938).
[95] P. H. Van Cittert, Die wahrscheinliche Schwingungsverteilung in einer von einer Lichtquelle direkt oder mittels einer Linse beleuchteten Ebene, *Physica* **1**, 201 (1934).
[96] P. H. Van Cittert, Kohaerenz-probleme, *Physica* **6**, 1129 (1939).

[97] M. von Laue, Die Entropie von partiell kohärenten Strahlenbündeln, *Ann. Phys.* **23**, 1 (1907).
[98] H. H. Hopkins, The concept of partial coherence in optics, *Proc. Roy. Soc. A* **208**, 263 (1951).
[99] H. H. Hopkins, On the diffraction theory of optical images, *Proc. Roy. Soc. A* **217**, 408 (1953).
[100] L. Mandel and E. Wolf, Coherence properties of optical fields, *Rev. Mod. Phys.* **37**, 231, 1965.
[101] R. E. Swing and J. R. Clay, Ambiguity of the transfer function with partially coherent illumination, *J. Opt. Soc. Am.* **57**, 1180 (1967).
[102] R. E. Kinzly, Partially coherent imaging in a microdensitometer, *J. Opt. Soc. Am.* **72**, 386 (1972).
[103] R. J. Becherer and G. B. Parrent, Jr., Nonlinearity in optical imaging systems, *J. Opt. Soc. Am.* **57**, 1479 (1967).
[104] R. N. Bracewell, "The Fourier Transform and its Applications." McGraw-Hill, New York, 1965.
[105] B. J. Thompson, Image formation with partially coherent light, *Progr. Opt.* **7**, 169 (1969).
[106] A. Walters, "Applied Optics and Optical Engineering" (R. Kingslake, ed.), Vol. I, p. 245. Academic Press, New York, 1965.
[107] J. DeVelis and G. B. Parrent, Transfer function for cascaded optical systems, *J. Opt. Soc. Am.* **57**, 1486 (1967).
[108] D. Grimes and B. J. Thompson, Two-point resolution with partially coherent light, *J. Opt. Soc. Am.* **57**, 1330 (1967).
[109] F. J. Zucker, "Introduction to Partially Coherent Electromagnetic Waves" (*Proc. Symp. Electromagn. Theory Antennas, Copenhagen*). Macmillan, New York, 1963.
[110] R. Hanbury Brown, R. C. Jennison, and M. K. Das Gupta, Apparent angular sizes of discrete radio sources, *Nature* (*London*) **170**, 1061 (1952).
[111] R. Hanbury Brown and R. Q. Twiss, A new type of interferometer for use in radio astronomy, *Phil. Mag. Ser. 7* **45**, 663 (1954).
[112] R. Hanbury Brown and R. Q. Twiss, Correlation between photons in two coherent beams of light, *Nature* (*London*) **177**, 27 (1956).
[113] R. Hanbury Brown and R. Q. Twiss, A test of a new type of stellar interferometer on sirius, *Nature* (*London*) **178**, 1046 (1956).
[114] R. Hanbury Brown and R. Q. Twiss, Interferometry of the intensity fluctuations in light, I. Basic theory: The correlation photons in coherent beams of radiation, *Proc. Roy. Soc.* **A242**, 300 (1957).
[115] R. Hanbury Brown and R. Q. Twiss, Interferometry of the intensity fluctuations in light, II. An experimental test of the theory for partially coherent light, *Proc. Roy. Soc.* **A243**, 291 (1957).
[116] R. Hanbury Brown and R. Q. Twiss, Interferometry of the intensity fluctuations in light, III. Applications to astronomy, *Proc. Roy. Soc.* **A248**, 199 (1958).
[117] R. Hanbury Brown and R. Q. Twiss, Interferometry of the intensity fluctuations in light, IV. A test of an intensity interferometer on sirius A, *Proc. Roy. Soc.* **A248**, 222 (1958).
[118] M. Harwit, Measurement of thermal fluctuations in radiation, *Phys. Rev.* **120**, 1551 (1960).

[119] R. Hanbury Brown, *Proc. Symp. Interferometry, N.P.L., London* (1960).
[120] J. W. Strohbehn, Line-of-sight wave propagation through the turbulent atmosphere, *Proc. IEEE* **56**, 1301 (1968).
[121] L. Mandel, Fluctuations of light beams, *Progr. Opt.* **2**, 183 (1968).
[122] R. E. Swing, Conditions for microdensitometer linearity, *J. Opt. Soc. Am.* **62**, 199 (1972).
[123] J. C. Dainty, Diffraction-limited imaging of stellar objects using telescopes of low optical quality, *Opt. Commun.* **7**, 129 (1973).
[124] Lord Rayleigh, "The Scientific Papers." Dover, New York, 1964; *Appl. Opt.* October (1964) (Special issue).
[125] H. C. Van de Hulst, "Light Scattering by Small Particles." Wiley, New York, 1957.
[126] J. W. Strohbehn, Optical propagation through the turbulent atmosphere, *Progr. Opt.* **9**, 75 (1971).
[127] S. Chandrasekhar, "Radiative Transfer." Dover, New York, 1960.
[128] A. Ishimaru, Correlation functions of a wave in a random distribution of stationary and moving scatterers, *Radio Sci.* **10**, 45 (1975).
[129] A. Ishimaru and S. T. Hong, Multiple scattering effects on coherent bandwidth and pulse distortion of a wave propagating in a random distribution of particles, *Radio Sci.* **10**, 637 (1975).
[130] A. G. Borovoy, M. V. Kabanov, and B. A. Saveliev, Intensity fluctuations of optical radiation in scattering media, *Appl. Opt.* **14**, 2731 (1975).
[131] K. Bullrich, Scattered radiation in the atmosphere and the natural aerosol, *Advan. Geophys.* **10**, 99 (1964).
[132] G. Mie, *Ann. Phys.* **25**, 377 (1908).
[133] P. Debye, *Ann. Phys.* (4), **30**, 57 (1909).
[134] P. Barber and C. Yeh, Scattering of electromagnetic waves by arbitrarily shaped dielectric bodies, *Appl. Opt.* **14**, 2864 (1975).
[135] C. I. Beard, T. H. Kays, and V. Twersky, Scattered intensities for random distributions—Microwave data and optical applications, *Appl. Opt.* **4**, 1299 (1965).
[136] R. Penndorf, *J. Phys. Chem.* **62**, 1537 (1958).
[137] V. Twersky, "On Propagation in Random Media of Discrete Scatterers" (*Proc. Symp. Appl. Math.*), Vol. XVI. Amer. Math. Soc., 1964.
[138] L. A. Chernov, "Wave Propagation in a Random Media." McGraw-Hill, New York, 1960.
[139] R. L. Fante, Electromagnetic beam propagation in turbulent media, *Proc. IEEE* **63**, 1669 (1975).
[140] H. B. Phillips, "Vector Analysis." Wiley, New York, 1960.
[141] H. S. Shapiro and R. A. Silverman, Some spectral properties of weighted random processes, *IRE Trans. Infor. Theor.* **IT-5**, 123 (1959).
[142] V. Tatarski, "The Effect of the Turbulent Atmosphere on Wave Propagation," Russian 1967; Springfield, Virginia, U.S. Dept. of Commerce, 1971 (Translation).
[143] P. Deitz and N. Wright, Saturation of scintillation magnitude in near-earth optical propagation, *J. Opt. Soc. Am.* **59**, 527 (1969).
[144] F. P. Carlson and A. Ihsimaru, Spherical waves in locally homogeneous media, *J. Opt. Soc. Am.* **59**, 319 (1969).
[145] A. Ishimaru, Fluctuations of a beam propagating through a locally homogeneous medium, *Radio Sci.* **4**, 295 (1969).

[146] A. Ishimaru, Fluctuations of a focused beam wave for atmospheric turbulence probing, *Proc. IEEE* **57**, 407 (1969).

[147] D. A. deWolf, Saturation of irradiance fluctuations due to turbulent atmosphere, *J. Opt. Soc. Am.* **58**, 461 (1968).

[148] S. F. Clifford, G. R. Ochs, and R. S. Lawrence, Saturation of optical scintillation by strong turbulence, *J. Opt. Soc. Am.* **64**, 148 (1974).

[149] M. I. Sancer and A. D. Varvatsis, Saturation calculation for light propagation in the turbulent atmosphere, *J. Opt. Soc. Am.* **60**, 654 (1970).

[150] W. P. Brown, Jr., Calculation of the variance of irradiance scintillation, *J. Opt. Soc. Am.* **61**, 981 (1971).

[151] U. Frisch, Wave propagation in random media, "Probabilistic Methods in Applied Mathematics" (A. T. Bharucha-Reid, ed.), Vol. 1. Academic Press, New York, 1968.

[152] D. A. deWolf, Strong irradiance fluctuations in turbulent air, I: Plane waves, *J. Opt. Soc. Am.* **63**, 171 (1973).

[153] D. A. deWolf, Strong irradiance fluctuations in turbulent air, II. Spherical waves, *J. Opt. Soc. Am.* **63**, 1249 (1973).

[154] A. S. Gurvich and V. I. Tatarski, Coherence and intensity fluctuations of light in the turbulent atmosphere, *Radio Sci.* **10**, 3 (1975).

[155] H. Yura, Physical model for strong optical-amplitude fluctuations in a turbulent medium, *J. Opt. Soc. Am.* **64**, 59 (1974).

[156] L. L. Foldy, The multiple scattering of waves, *Phys. Rev.* **67**, 107 (1945).

[157] M. Lax, Multiple scattering of waves, *Rev. Mod. Phys.* **23**, 287 (1951).

[158] J. E. Burke and V. Twersky, On scattering of waves by many bodies, *J. Res. Nat. Bur. Std. Sect. D* **68D**, 500 (1964).

[159] R. Fante, Propagation of electromagnetic waves through turbulent plasma using transport theory, *IEEE Trans. Antennas Propagation* **AP-21**, 750 (1973).

[160] R. Fante, Mutual coherence function and frequency spectrum of a laser beam propagating through atmospheric turbulence, *J. Opt. Soc. Am.* **64**, 592 (1974).

[161] B. Davison, "Neutron Transport Theory." Oxford Univ. Press, London and New York, 1958.

[162] A. Ishimaru and J. C. Lin, Multiple scattering effects on wave propagation through rain, *AGARD Conf. Proc. No. 107* **1-1**, April 1973.

[163] K. Furutsu, Multiple scattering of waves in a medium of randomly distributed particles and derivation of the transport equation, *Radio Sci.* **10**, 29 (1975).

[164] E. A. Bucher and R. M. Lerner, Experiments on light pulse communication through thick clouds, *Appl. Opt.* **12**, 1201 (1973).

[165] A. Ishimaru, "Wave Propagation and Scattering in Random Media." Academic Press, New York, 1977.

INDEX

B

C

G

H

N

O

P

S

A B 7 C 8 D 9 E 0 F 1 G 2 H 3 I 4 J 5